The Descent of Man

By
Charles Darwin

Shine Classics

ISBN: 978-1503305908

INTRODUCTION

The nature of the following work will be best understood by a brief account of how it came to be written. During many years I collected notes on the origin or descent of man, without any intention of publishing on the subject, but rather with the determination not to publish, as I thought that I should thus only add to the prejudices against my views. It seemed to me sufficient to indicate, in the first edition of my 'Origin of Species,' that by this work "light would be thrown on the origin of man and his history;" and this implies that man must be included with other organic beings in any general conclusion respecting his manner of appearance on this earth. Now the case wears a wholly different aspect. When a naturalist like Carl Vogt ventures to say in his address as President of the National Institution of Geneva (1869), "personne, en Europe au moins, n'ose plus soutenir la création indépendante et de toutes pièces, des espèces," it is manifest that at least a large number of naturalists must admit that species are the modified descendants of other species;2 and this especially holds good with the younger and rising naturalists. The greater number accept the agency of natural selection; though some urge, whether with justice the future must decide, that I have greatly overrated its importance. Of the older and honoured chiefs in natural science, many unfortunately are still opposed to evolution in every form.

In consequence of the views now adopted by most naturalists, and which will ultimately, as in every other case, be followed by other men, I have been led to put together my notes, so as to see how far the general conclusions arrived at in my former works were applicable to man. This seemed all the more desirable as I had never deliberately applied these views to a species taken singly. When we confine our attention to any one form, we are deprived of the weighty arguments derived from the nature of the affinities which connect together whole groups of organisms—their geographical distribution in past and present times, and their geological succession. The homological structure, embryological development, and rudimentary organs of a species, whether it be man or any other animal, to which our attention may be directed, remain to be considered; but these great classes of facts afford, as it appears to me, ample and conclusive evidence in favour of the principle of gradual evolution. The strong support derived from the other arguments should, however, always be kept before the mind.

The sole object of this work is to consider, firstly, whether man, like every other species, is descended from some pre-existing form; secondly, the manner of3 his development; and thirdly, the value of the differences between the so-called races of man. As I shall confine myself to these points, it will not be necessary to describe in detail the differences between the several races—an enormous subject which has been fully discussed in many valuable works. The high antiquity of man has recently been demonstrated by the labours of a host of eminent men, beginning with M. Boucher de Perthes; and this is the indispensable basis for understanding his origin. I shall, therefore, take this conclusion for granted, and may refer my readers to the admirable treatises of Sir Charles Lyell, Sir John Lubbock, and others. Nor shall I have occasion to do more than to allude to the amount of difference between man and the anthropomorphous apes; for Prof. Huxley, in the opinion of most competent judges, has conclusively shewn that in every single visible character man differs less from the higher apes than these do from the lower members of the same order of Primates.

This work contains hardly any original facts in regard to man; but as the conclusions at which I arrived, after drawing up a rough draft, appeared to me interesting, I thought that they might interest others. It has often and confidently been asserted, that man's origin can never be known: but ignorance more frequently begets confidence than does knowledge: it is those who know little, and not those who know much, who so positively assert that this or that problem will never be solved by science. The conclusion that man is the co-descendant with other species of some ancient, lower, and extinct form, is not in any degree new. La4marck long ago came to this conclusion, which has lately been maintained by several eminent naturalists and philosophers; for instance by Wallace, Huxley, Lyell, Vogt, Lubbock, Büchner, Rolle, &c.,1 and especially by Häckel. This last naturalist, besides his great work, 'Generelle Morphologie '(1866), has recently (1868, with a second edit. in 1870), published his 'Natürliche Schöpfungsgeschichte, 'in which he fully discusses the genealogy of man. If this work had appeared before my essay had been written, I should probably never have completed it. Almost all the conclusions at which I have arrived I find confirmed by this naturalist, whose knowledge on many points is much fuller than mine. Wherever I have added any fact or view from Prof. Häckel's writings, I give his authority in the text, other statements I leave as they originally stood in my manuscript, occasionally giving in the foot-notes references to his works, as a confirmation of the more doubtful or interesting points.

During many years it has seemed to me highly probable that sexual selection has played an important part in differentiating the races of man; but in my 5 6'Origin of Species' (first edition, p. 199) I contented myself by merely alluding to this belief. When I came to apply this view to man, I found it indispensable to treat the whole subject in full detail.2Consequently the second part of the present work, treating of sexual selection, has extended to an inordinate length, compared with the first part; but this could not be avoided.

I had intended adding to the present volumes an essay on the expression of the various emotions by man and the lower animals. My attention was called to this subject many years ago by Sir Charles Bell's admirable work. This illustrious anatomist maintains that man is endowed with certain muscles solely for the sake of expressing his emotions. As this view is obviously opposed to the belief that man is descended from some other and lower form, it was necessary for me to consider it. I likewise wished to ascertain how far the emotions are expressed in the same manner by the different races of man. But owing to the length of the present work, I have thought it better to reserve my essay, which is partially completed, for separate publication.

Part I.
THE DESCENT OR ORIGIN OF MAN.

CHAPTER I.
The Evidence of the Descent of Man from some Lower Form.

Nature of the evidence bearing on the origin of man—Homologous structures in man and the lower animals—Miscellaneous points of correspondence—Development—Rudimentary structures, muscles, sense-organs, hair, bones, reproductive organs, &c.—The bearing of these three great classes of facts on the origin of man.

He who wishes to decide whether man is the modified descendant of some pre-existing form, would probably first enquire whether man varies, however slightly, in bodily structure and in mental faculties; and if so, whether the variations are transmitted to his offspring in accordance with the laws which prevail with the lower animals; such as that of the transmission of characters to the same age or sex. Again, are the variations the result, as far as our ignorance permits us to judge, of the same general causes, and are they governed by the same general laws, as in the case of other organisms; for instance by correlation, the inherited effects of use and disuse, &c.? Is man subject to similar malconformations, the result of arrested development, of reduplication of parts, &c., and does he display in any of his anomalies reversion to some former and ancient type of structure? It might also naturally be enquired whether man, like so many other animals, has given rise to varieties and sub-races, differing but slightly from each other, or to 10 races differing so much that they must be classed as doubtful species? How are such races distributed over the world; and how, when crossed, do they react on each other, both in the first and succeeding generations? And so with many other points.

The enquirer would next come to the important point, whether man tends to increase at so rapid a rate, as to lead to occasional severe struggles for existence, and consequently to beneficial variations, whether in body or mind, being preserved, and injurious ones eliminated. Do the races or species of men, whichever term may be applied, encroach on and replace each other, so that some finally become extinct? We shall see that all these questions, as indeed is obvious in respect to most of them, must be answered in the affirmative, in the same manner as with the lower animals. But the several considerations just referred to may be conveniently deferred for a time; and we will first see how far the bodily structure of man shows traces, more or less plain, of his descent from some lower form. In the two succeeding chapters the mental powers of man, in comparison with those of the lower animals, will be considered.

The Bodily Structure of Man.—It is notorious that man is constructed on the same general type or model with other mammals. All the bones in his skeleton can be compared with corresponding bones in a monkey, bat, or seal. So it is with his muscles, nerves, blood-vessels and internal viscera. The brain, the most important of all the organs, follows the same law, as shewn by Huxley and other anatomists. Bischoff,[3] who is a hostile witness, admits that every chief fissure and fold 11 in the brain of man has its analogy in that of the orang; but he adds that at no period of development do their brains perfectly agree; nor could this be expected, for otherwise their mental powers would have been the same. Vulpian[4] remarks: "Les différences réelles qui existent entre l'encéphale de l'homme et celui des singes supérieurs, sont bien minimes. Il ne faut pas se faire d'illusions à cet égard.

L'homme est bien plus près des singes anthropomorphes par les caractères anatomiques de son cerveau que ceux-ci ne le sont non-seulement des autres mammifères, mais mêmes de certains quadrumanes, des guenons et des macaques." But it would be superfluous here to give further details on the correspondence between man and the higher mammals in the structure of the brain and all other parts of the body.

It may, however, be worth while to specify a few points, not directly or obviously connected with structure, by which this correspondence or relationship is well shewn.

Man is liable to receive from the lower animals, and to communicate to them, certain diseases as hydrophobia, variola, the glanders, &c.; and this fact proves the close similarity of their tissues and blood, both in minute structure and composition, far more plainly than does their comparison under the best microscope, or by the aid of the best chemical analysis. Monkeys are liable to many of the same non-contagious diseases as we are; thus Rengger,[5] who carefully observed for a long time the Cebus Azaræ in its native land, found it liable to catarrh, with the usual symptoms, and which when 12often recurrent led to consumption. These monkeys suffered also from apoplexy, inflammation of the bowels, and cataract in the eye. The younger ones when shedding their milk-teeth often died from fever. Medicines produced the same effect on them as on us. Many kinds of monkeys have a strong taste for tea, coffee, and spirituous liquors: they will also, as I have myself seen, smoke tobacco with pleasure. Brehm asserts that the natives of north-eastern Africa catch the wild baboons by exposing vessels with strong beer, by which they are made drunk. He has seen some of these animals, which he kept in confinement, in this state; and he gives a laughable account of their behaviour and strange grimaces. On the following morning they were very cross and dismal; they held their aching heads with both hands and wore a most pitiable expression: when beer or wine was offered them, they turned away with disgust, but relished the juice of lemons.[6] An American monkey, an Ateles, after getting drunk on brandy, would never touch it again, and thus was wiser than many men. These trifling facts prove how similar the nerves of taste must be in monkeys and man, and how similarly their whole nervous system is affected.

Man is infested with internal parasites, sometimes causing fatal effects, and is plagued by external parasites, all of which belong to the same genera or families with those infesting other mammals. Man is subject like other mammals, birds, and even insects, to that mysterious law, which causes certain normal processes, such as gestation, as well as the maturation and duration of various diseases, to follow lunar periods.[7] His wounds 13are repaired by the same process of healing; and the stumps left after the amputation of his limbs occasionally possess, especially during an early embryonic period, some power of regeneration, as in the lowest animals.[8]

The whole process of that most important function, the reproduction of the species, is strikingly the same in all mammals, from the first act of courtship by the male[9] to the birth and nurturing of the young. Monkeys are born in almost as helpless a condition as our own infants; and in certain genera the young differ fully as much in appearance from the adults, as do our children from their full-grown parents.[10] It has been urged by some writers as an important distinction, that with man the young arrive at maturity at a much later age than with any other animal; but if we look to the races of mankind which inhabit tropical countries the difference is not great, for the orang is believed not to be adult till the age of from ten to fifteen years.[11] Man 14differs from woman in size, bodily strength, hairyness, &c., as well as in mind, in the same manner as do the two sexes of many mammals. It is, in short, scarcely possible to exaggerate the close correspondence in general structure, in the minute structure of the tissues, in chemical composition and in constitution, between man and the higher animals, especially the anthropomorphous apes.

Embryonic Development.—Man is developed from an ovule, about the 125th of an inch in diameter, which differs in no respect from the ovules of other animals. The embryo itself at a very early period can hardly be distinguished from that of other members of the vertebrate kingdom. At this period the arteries run in arch-like branches, as if to carry the blood to branchiæ which are not present in the higher vertebrata, though the slits on the sides of the neck still remain (f, g, fig. 1), marking their former position. At a somewhat later period, when the extremities are developed, "the feet of lizards and mammals," as the illustrious Von Baer remarks, "the wings and feet of birds, no less than the hands and feet of man, all arise from the same fundamental form." It is, says Prof. Huxley,[12] "quite in the later stages of development that the young human being presents marked differences from the young ape, while the latter departs as much from the dog in its developments, as the man does. Startling as this last assertion may appear to be, it is demonstrably true."

As some of my readers may never have seen a drawing of an embryo, I have given one of man and another of a dog, at about the same early stage of development, [15]carefully copied from two works of undoubted accuracy.[13]

After the foregoing statements made by such high authorities, it would be superfluous on my part to give a number of borrowed details, shewing that the embryo of man closely resembles that of other mammals. It may, however, be added that the human embryo likewise resembles in various points of structure certain low forms when adult. For instance, the heart at first exists as a simple pulsating vessel; the excreta are voided through a cloacal passage; and the os coccyx projects like a true tail, "extending considerably beyond the rudimentary legs."[14] In the embryos of all air-breathing vertebrates, certain glands called the corpora Wolffiana, correspond with and act like the kidneys of mature fishes.[15] Even at a later embryonic period, some striking resemblances between man and the lower animals may be observed. Bischoff says that the convolutions of the brain in a human fœtus at the end of the seventh month reach about the same stage of development as in a baboon when adult.[16] The great toe, as Prof. Owen remarks,[17] "which forms the fulcrum when standing or walking, is perhaps the most characteristic [17]peculiarity in the human structure;" but in an embryo, about an inch in length, Prof. Wyman[18] found "that the great toe was shorter than the others, and, instead of being parallel to them, projected at an angle from the side of the foot, thus corresponding with the permanent condition of this part in the quadrumana." I will conclude with a quotation from Huxley,[19] who after asking, does man originate in a different way from a dog, bird, frog or fish? says, "the reply is not doubtful for a moment; without question, the mode of origin and the early stages of the development of man are identical with those of the animals immediately below him in the scale: without a doubt in these respects, he is far nearer to apes, than the apes are to the dog."

Rudiments.—This subject, though not intrinsically more important than the two last, will for several reasons be here treated with more fullness.[20] Not one of the higher animals can be named which does not bear some part in a rudimentary condition; and man forms no exception to the rule. Rudimentary organs must be distinguished from those that are nascent; though in some cases the distinction is not easy. The former are either absolutely useless, such as the mammæ of male quadrupeds, or the incisor teeth of ruminants which never cut through the gums; or they are of such slight service to their present possessors, that we cannot suppose that they were developed under the conditions [18]which now exist.

Organs in this latter state are not strictly rudimentary, but they are tending in this direction. Nascent organs, on the other hand, though not fully developed, are of high service to their possessors, and are capable of further development. Rudimentary organs are eminently variable; and this is partly intelligible, as they are useless or nearly useless, and consequently are no longer subjected to natural selection. They often become wholly suppressed. When this occurs, they are nevertheless liable to occasional reappearance through reversion; and this is a circumstance well worthy of attention.

Disuse at that period of life, when an organ is chiefly used, and this is generally during maturity, together with inheritance at a corresponding period of life, seem to have been the chief agents in causing organs to become rudimentary. The term "disuse" does not relate merely to the lessened action of muscles, but includes a diminished flow of blood to a part or organ, from being subjected to fewer alternations of pressure, or from becoming in any way less habitually active. Rudiments, however, may occur in one sex of parts normally present in the other sex; and such rudiments, as we shall hereafter see, have often originated in a distinct manner. In some cases organs have been reduced by means of natural selection, from having become injurious to the species under changed habits of life. The process of reduction is probably often aided through the two principles of compensation and economy of growth; but the later stages of reduction, after disuse has done all that can fairly be attributed to it, and when the saving to be effected by the economy of growth would be very small,[21] are difficult to understand. The final and com19plete suppression of a part, already useless and much reduced in size, in which case neither compensation nor economy can come into play, is perhaps intelligible by the aid of the hypothesis of pangenesis, and apparently in no other way. But as the whole subject of rudimentary organs has been fully discussed and illustrated in my former works,[22] I need here say no more on this head.

Rudiments of various muscles have been observed in many parts of the human body;[23] and not a few muscles, which are regularly present in some of the lower animals can occasionally be detected in man in a greatly reduced condition. Every one must have noticed the power which many animals, especially horses, possess of moving or twitching their skin; and this is effected by the panniculus carnosus. Remnants of this muscle in an efficient state are found in various parts of our bodies; for instance, on the forehead, by which the eyebrows are raised. The platysma myoides, which is well developed on the neck, belongs to this system, but cannot be voluntarily brought into action. Prof. Turner, of Edinburgh, has occasionally detected, as he informs me, muscular fasciculi in five different situations, namely in the axillæ, near the scapulæ, &c., all of which must be referred to the system of the panniculus. He has also shewn[24] that the musculus sternalis or sternalis brutorum, which is not an extension of the rectus abdominalis, but is closely allied to the panniculus, oc20curred in the proportion of about 3 per cent. in upwards of 600 bodies: he adds, that this muscle affords "an excellent illustration of the statement that occasional and rudimentary structures are especially liable to variation in arrangement."

Some few persons have the power of contracting the superficial muscles on their scalps; and these muscles are in a variable and partially rudimentary condition. M. A. de Candolle has communicated to me a curious instance of the long-continued persistence or inheritance of this power, as well as of its unusual development. He knows a family, in which one member, the present head of a family, could, when a youth, pitch several heavy books from his head by the movement of the scalp alone; and he won wagers by performing this feat. His father, uncle, grandfather, and all his three children possess the same power to the same unusual degree. This family became divided eight generations ago into two branches; so that the head of the above-mentioned branch is cousin in the seventh degree to the head of the other branch. This distant cousin resides in another part of France, and on being asked whether he possessed the same faculty, immediately exhibited his power. This case offers a good illustration how persistently an absolutely useless faculty may be transmitted.

The extrinsic muscles which serve to move the whole external ear, and the intrinsic muscles which move the different parts, all of which belong to the system of the panniculus, are in a rudimentary condition in man; they are also variable in development, or at least in function. I have seen one man who could draw his ears forwards, and another who could draw them backwards;[25]

21and from what one of these persons told me, it is probable that most of us by often touching our ears and thus directing our attention towards them, could by repeated trials recover some power of movement. The faculty of erecting the ears and of directing them to different points of the compass, is no doubt of the highest service to many animals, as they thus perceive the point of danger; but I have never heard of a man who possessed the least power of erecting his ears,—the one movement which might be of use to him. The whole external shell of the ear may be considered a rudiment, together with the various folds and prominences (helix and anti-helix, tragus and anti-tragus, &c.) which in the lower animals strengthen and support the ear when erect, without adding much to its weight. Some authors, however, suppose that the cartilage of the shell serves to transmit vibrations to the acoustic nerve; but Mr. Toynbee,[26] after collecting all the known evidence on this head, concludes that the external shell is of no distinct use. The ears of the chimpanzee and orang are curiously like those of man, and I am assured by the keepers in the Zoological Gardens that these animals never move or erect them; so that they are in an equally rudimentary condition, as far as function is concerned, as in man. Why these animals, as well as the progenitors of man, should have lost the power of erecting their ears we cannot say. It may be, though I am not quite satisfied with this view, that owing to their arboreal habits and great strength they were but little exposed to danger, and so during a lengthened period moved their ears but little, and thus gradually lost the power of moving them. This would be a parallel case with that of those large and heavy birds, 22which from inhabiting oceanic islands have not been exposed to the attacks of beasts of prey, and have consequently lost the power of using their wings for flight.

The celebrated sculptor, Mr. Woolner, informs me of one little peculiarity in the external ear, which he has often observed both in men and women, and of which he perceived the full signification. His attention was first called to the subject whilst at work on his figure of Puck, to which he had given pointed ears. He was thus led to examine the ears of various monkeys, and subsequently more carefully those of man. The peculiarity consists in a little blunt point, projecting from the inwardly folded margin, or helix. Mr. Woolner made an exact model of one such case, and has sent me the accompanying drawing. (Fig. 2.) These points not only project inwards, but often a little outwards, so that they are visible when the head is viewed from directly in front or behind. They are variable in size and somewhat in position, standing either a little higher or lower; and they sometimes occur on one ear and not on the other. Now the meaning of these projections is not, I think, doubtful; but it may be thought that they offer too trifling a character to be worth notice. This thought, however, is as false as it is natural. Every character, however slight, must be the result of some definite cause; and if it occurs in many individuals deserves consideration. The helix obviously consists of the extreme margin of the ear folded inwards; and this folding appears to be in some manner connected with the23 whole external ear being permanently pressed backwards. In many monkeys, which do not stand high in the order, as baboons and some species of macacus,27 the upper portion of the ear is slightly pointed, and the margin is not at all folded inwards; but if the margin were to be thus folded, a slight point would necessarily project inwards and probably a little outwards. This could actually be observed in a specimen of the Ateles beelzebuth in the Zoological Gardens; and we may safely conclude that it is a similar structure—a vestige of formerly pointed ears—which occasionally reappears in man.

The nictitating membrane, or third eyelid, with its accessory muscles and other structures, is especially well developed in birds, and is of much functional importance to them, as it can be rapidly drawn across the whole eye-ball. It is found in some reptiles and amphibians, and in certain fishes, as in sharks. It is fairly well developed in the two lower divisions of the mammalian series, namely, in the monotremata and marsupials, and in some few of the higher mammals, as in the walrus. But in man, the quadrumana, and most other mammals, it exists, as is admitted by all anatomists, as a mere rudiment, called the semilunar fold.28

The sense of smell is of the highest importance to the greater number of mammals—to some, as the ruminants, in warning them of danger; to others, as the 24carnivora, in finding their prey; to others, as the wild boar, for both purposes combined. But the sense of smell is of extremely slight service, if any, even to savages, in whom it is generally more highly developed than in the civilised races. It does not warn them of danger, nor guide them to their food; nor does it prevent the Esquimaux from sleeping in the most fetid atmosphere, nor many savages from eating half-putrid meat. Those who believe in the principle of gradual evolution, will not readily admit that this sense in its present state was originally acquired by man, as he now exists. No doubt he inherits the power in an enfeebled and so far rudimentary condition, from some early progenitor, to whom it was highly serviceable and by whom it was continually used. We can thus perhaps understand how it is, as Dr. Maudsley has truly remarked,29 that the sense of smell in man "is singularly effective in recalling vividly the ideas and images of forgotten scenes and places;" for we see in those animals, which have this sense highly developed, such as dogs and horses, that old recollections of persons and places are strongly associated with their odour.

Man differs conspicuously from all the other Primates in being almost naked. But a few short straggling hairs are found over the greater part of the body in the male sex, and fine down on that of the female sex, In individuals belonging to the same race these hairs are highly variable, not only in abundance, but likewise in position: thus the shoulders in some Europeans are quite naked, whilst in others they bear thick tufts of hair.30 There can be little doubt that the hairs 25thus scattered over the body are the rudiments of the uniform hairy coat of the lower animals. This view is rendered all the more probable, as it is known that fine, short, and pale-coloured hairs on the limbs and other parts of the body occasionally become developed into "thickset, long, and rather coarse dark hairs," when abnormally nourished near old-standing inflamed surfaces.31

I am informed by Mr. Paget that persons belonging to the same family often have a few hairs in their eyebrows much longer than the others; so that this slight peculiarity seems to be inherited. These hairs apparently represent the vibrissæ, which are used as organs of touch by many of the lower animals. In a young chimpanzee I observed that a few upright, rather long, hairs, projected above the eyes, where the true eyebrows, if present, would have stood.

The fine wool-like hair, or so-called lanugo, with which the human fœtus during the sixth month is thickly covered, offers a more curious case. It is first developed, during the fifth month, on the eyebrows and face, and especially round the mouth, where it is much longer than that on the head. A moustache of this kind was observed by Eschricht32 on a female fœtus; but this is not so surprising a circumstance as it may at first appear, for the two sexes generally resemble each other in all external characters during an early period of growth. The direction and arrangement of the hairs on all parts of the fœtal body are the same as in the adult, but are subject to much variability. The whole surface, including even the forehead and ears, is thus thickly clothed; but it is a significant fact that the palms of the hands and 26the soles of the feet are quite naked, like the inferior surfaces of all four extremities in most of the lower animals. As this can hardly be an accidental coincidence, we must consider the woolly covering of the fœtus to be the rudimental representative of the first permanent coat of hair in those mammals which are born hairy. This representation is much more complete, in accordance with the usual law of embryological development, than that afforded by the straggling hairs on the body of the adult.

It appears as if the posterior molar or wisdom-teeth were tending to become rudimentary in the more civilised races of man. These teeth are rather smaller than the other molars, as is likewise the case with the corresponding teeth in the chimpanzee and orang; and they have only two separate fangs. They do not cut through the gums till about the seventeenth year, and I am assured by dentists that they are much more liable to decay, and are earlier lost, than the other teeth. It is also remarkable that they are much more liable to vary both in structure and in the period of their development than the other teeth.33 In the Melanian races, on the other hand, the wisdom-teeth are usually furnished with three separate fangs, and are generally sound: they also differ from the other molars in size less than in the Caucasian races.34 Prof. Schaaffhausen accounts for this difference between the races by "the posterior dental portion of the jaw being always shortened" in those that are civilised,35 and this shortening may, I presume, be safely attributed to civi27lised men habitually feeding on soft, cooked food, and thus using their jaws less. I am informed by Mr. Brace that it is becoming quite a common practice in the United States to remove some of the molar teeth of children, as the jaw does not grow large enough for the perfect development of the normal number.

With respect to the alimentary canal I have met with an account of only a single rudiment, namely the vermiform appendage of the cæcum. The cæcum is a branch or diverticulum of the intestine, ending in a cul-de-sac, and it is extremely long in many of the lower vegetable-feeding mammals. In the marsupial koala it is actually more than thrice as long as the whole body.36 It is sometimes produced into a long gradually-tapering point, and is sometimes constricted in parts. It appears as if, in consequence of changed diet or habits, the cæcum had become much shortened in various animals, the vermiform appendage being left as a rudiment of the shortened part. That this appendage is a rudiment, we may infer from its small size, and from the evidence which Prof. Canestrini37 has collected of its variability in man. It is occasionally quite absent, or again is largely developed. The passage is sometimes completely closed for half or two-thirds of its length, with the terminal part consisting of a flattened solid expansion. In the orang this appendage is long and convoluted: in man it arises from the end of the short cæcum, and is commonly from four to five inches in length, being only about the third of an inch in diameter. Not only is it useless, but it is sometimes the cause of death, of which fact I have lately heard two instances: this is due to small hard bodies, 28such as seeds, entering the passage and causing inflammation.38

In the Quadrumana and some other orders of mammals, especially in the Carnivora, there is a passage near the lower end of the humerus, called the supra-condyloid foramen, through which the great nerve of the fore limb passes, and often the great artery. Now in the humerus of man, as Dr. Struthers[39] and others have shewn, there is generally a trace of this passage, and it is sometimes fairly well developed, being formed by a depending hook-like process of bone, completed by a band of ligament. When present the great nerve invariably passes through it, and this clearly indicates that it is the homologue and rudiment of the supra-condyloid foramen of the lower animals. Prof. Turner estimates, as he informs me, that it occurs in about one per cent. of recent skeletons; but during ancient times it appears to have been much more common. Mr. Busk[40] has collected the following evidence on this head: Prof. Broca "noticed the perforation in four and a half per cent. of the arm-bones collected in the 'Cimetière du Sud' at Paris; and in the Grotto of Orrony, the contents of which are referred to the Bronze period, as many as eight humeri out of thirty-two were perforated; but this extraordinary proportion, he thinks, might be due to the cavern having been a sort of 29'family vault.' Again, M. Dupont found 30 per cent. of perforated bones in the caves of the Valley of the Lesse, belonging to the Reindeer period; whilst M. Leguay, in a sort of dolmen at Argenteuil, observed twenty-five per cent. to be perforated; and M. Pruner-Bey found twenty-six per cent. in the same condition in bones from Vauréal. Nor should it be left unnoticed that M. Pruner-Bey states that this condition is common in Guanche skeletons." The fact that ancient races, in this and several other cases, more frequently present structures which resemble those of the lower animals than do the modern races, is interesting. One chief cause seems to be that ancient races stand somewhat nearer than modern races in the long line of descent to their remote animal-like progenitors.

The os coccyx in man, though functionless as a tail, plainly represents this part in other vertebrate animals. At an early embryonic period it is free, and, as we have seen, projects beyond the lower extremities. In certain rare and anomalous cases it has been known, according to Isidore Geoffroy St.-Hilaire and others,[41] to form a small external rudiment of a tail. The os coccyx is short, usually including only four vertebræ: and these are in a rudimental condition, for they consist, with the exception of the basal one, of the centrum alone.[42]They are furnished with some small muscles; one of which, as I am informed by Prof. Turner, has been expressly described by Theile as a rudimentary repetition of the extensor of the tail, which is so largely developed in many mammals.

The spinal cord in man extends only as far downwards as the last dorsal or first lumbar vertebra; but a 30thread-like structure (the filum terminale) runs down the axis of the sacral part of the spinal canal, and even along the back of the coccygeal bones. The upper part of this filament, as Prof. Turner informs me, is undoubtedly homologous with the spinal cord; but the lower part apparently consists merely of the pia mater, or vascular investing membrane. Even in this case the os coccyx may be said to possess a vestige of so important a structure as the spinal cord, though no longer enclosed within a bony canal. The following fact, for which I am also indebted to Prof. Turner, shews how closely the os coccyx corresponds with the true tail in the lower animals: Luschka has recently discovered at the extremity of the coccygeal bones a very peculiar convoluted body, which is continuous with the middle sacral artery; and this discovery led Krause and Meyer to examine the tail of a monkey (Macacus) and of a cat, in both of which they found, though not at the extremity, a similarly convoluted body.

The reproductive system offers various rudimentary structures; but these differ in one important respect from the foregoing cases. We are not here concerned with a vestige of a part which does not belong to the species in an efficient state; but with a part which is always present and efficient in the one sex, being represented in the other by a mere rudiment. Nevertheless, the occurrence of such rudiments is as difficult to explain on the belief of the separate creation of each species, as in the foregoing cases. Hereafter I shall have to recur to these rudiments, and shall shew that their presence generally depends merely on inheritance; namely, on parts acquired by one sex having been partially transmitted to the other. Here I will only give some instances of such rudiments. It is well known that in the males of all mammals, in31cluding man, rudimentary mammæ exist. These in several instances have become well developed, and have yielded a copious supply of milk. Their essential identity in the two sexes is likewise shewn by their occasional sympathetic enlargement in both during an attack of the measles. The vesicula prostratica, which has been observed in many male mammals, is now universally acknowledged to be the homologue of the female uterus, together with the connected passage. It is impossible to read Leuckart's able description of this organ, and his reasoning, without admitting the justness of his conclusion. This is especially clear in the case of those mammals in which the true female uterus bifurcates, for in the males of these the vesicula likewise bifurcates.43 Some additional rudimentary structures belonging to the reproductive system might here have been adduced.44

The bearing of the three great classes of facts now given is unmistakeable. But it would be superfluous here fully to recapitulate the line of argument given in detail in my 'Origin of Species.' The homological construction of the whole frame in the members of the same class is intelligible, if we admit their descent from a common progenitor, together with their subsequent adaptation to diversified conditions. On any other view the similarity of pattern between the hand of a man or monkey, the foot of a horse, the flipper of a seal, the wing of a bat, &c., is utterly inexplicable. It is no scientific 32explanation to assert that they have all been formed on the same ideal plan. With respect to development, we can clearly understand, on the principle of variations supervening at a rather late embryonic period, and being inherited at a corresponding period, how it is that the embryos of wonderfully different forms should still retain, more or less perfectly, the structure of their common progenitor. No other explanation has ever been given of the marvellous fact that the embryo of a man, dog, seal, bat, reptile, &c., can at first hardly be distinguished from each other. In order to understand the existence of rudimentary organs, we have only to suppose that a former progenitor possessed the parts in question in a perfect state, and that under changed habits of life they became greatly reduced, either from simple disuse, or through the natural selection of those individuals which were least encumbered with a superfluous part, aided by the other means previously indicated.

Thus we can understand how it has come to pass that man and all other vertebrate animals have been constructed on the same general model, why they pass through the same early stages of development, and why they retain certain rudiments in common. Consequently we ought frankly to admit their community of descent: to take any other view, is to admit that our own structure and that of all the animals around us, is a mere snare laid to entrap our judgment. This conclusion is greatly strengthened, if we look to the members of the whole animal series, and consider the evidence derived from their affinities or classification, their geographical distribution and geological succession. It is only our natural prejudice, and that arrogance which made our forefathers declare that they were descended from demi-gods, which leads us to demur to33 this conclusion. But the time will before long come when it will be thought wonderful, that naturalists, who were well acquainted with the comparative structure and development of man and other mammals, should have believed that each was the work of a separate act of creation.

CHAPTER II.

Comparison of the Mental Powers of Man and the Lower Animals.

The difference in mental power between the highest ape and the lowest savage, immense—Certain instincts in common—The emotions—Curiosity—Imitation—Attention—Memory—Imagination—Reason—Progressive improvement—Tools and weapons used by animals—Language—Self-consciousness—Sense of beauty—Belief in God, spiritual agencies, superstitions.

We have seen in the last chapter that man bears in his bodily structure clear traces of his descent from some lower form; but it may be urged that, as man differs so greatly in his mental power from all other animals, there must be some error in this conclusion. No doubt the difference in this respect is enormous, even if we compare the mind of one of the lowest savages, who has no words to express any number higher than four, and who uses no abstract terms for the commonest objects or affections,45 with that of the most highly organised ape. The difference would, no doubt, still remain immense, even if one of the higher apes had been improved or civilised as much as a dog has been in comparison with its parent-form, the wolf or jackal. The Fuegians rank amongst the lowest barbarians; but I was continually struck with surprise how closely the three natives on board H.M.S. "Beagle," who had lived some years in England and could talk a little English, resembled us in disposition and in most of our mental faculties. If no 35organic being excepting man had possessed any mental power, or if his powers had been of a wholly different nature from those of the lower animals, then we should never have been able to convince ourselves that our high faculties had been gradually developed. But it can be clearly shewn that there is no fundamental difference of this kind. We must also admit that there is a much wider interval in mental power between one of the lowest fishes, as a lamprey or lancelet, and one of the higher apes, than between an ape and man; yet this immense interval is filled up by numberless gradations.

Nor is the difference slight in moral disposition between a barbarian, such as the man described by the old navigator Byron, who dashed his child on the rocks for dropping a basket of sea-urchins, and a Howard or Clarkson; and in intellect, between a savage who does not use any abstract terms, and a Newton or Shakspeare. Differences of this kind between the highest men of the highest races and the lowest savages, are connected by the finest gradations. Therefore it is possible that they might pass and be developed into each other.

My object in this chapter is solely to shew that there is no fundamental difference between man and the higher mammals in their mental faculties. Each division of the subject might have been extended into a separate essay, but must here be treated briefly. As no classification of the mental powers has been universally accepted, I shall arrange my remarks in the order most convenient for my purpose; and will select those facts which have most struck me, with the hope that they may produce some effect on the reader.

With respect to animals very low in the scale, I shall have to give some additional facts under Sexual Selection, shewing that their mental powers are higher than36 might have been expected. The variability of the faculties in the individuals of the same species is an important point for us, and some few illustrations will here be given. But it would be superfluous to enter into many details on this head, for I have found on frequent enquiry, that it is the unanimous opinion of all those who have long attended to animals of many kinds, including birds, that the individuals differ greatly in every mental characteristic. In what manner the mental powers were first developed in the lowest organisms, is as hopeless an enquiry as how life first originated. These are problems for the distant future, if they are ever to be solved by man.

As man possesses the same senses with the lower animals, his fundamental intuitions must be the same. Man has also some few instincts in common, as that of self-preservation, sexual love, the love of the mother for her new-born offspring, the power possessed by the latter of sucking, and so forth. But man, perhaps, has somewhat fewer instincts than those possessed by the animals which come next to him in the series. The orang in the Eastern islands, and the chimpanzee in Africa, build platforms on which they sleep; and, as both species follow the same habit, it might be argued that this was due to instinct, but we cannot feel sure that it is not the result of both animals having similar wants and possessing similar powers of reasoning. These apes, as we may assume, avoid the many poisonous fruits of the tropics, and man has no such knowledge; but as our domestic animals, when taken to foreign lands and when first turned out in the spring, often eat poisonous herbs, which they afterwards avoid, we cannot feel sure that the apes do not learn from their own experience or from that of their parents what fruits to select. It is however certain, as we shall presently see, that apes have37 an instinctive dread of serpents, and probably of other dangerous animals.

The fewness and the comparative simplicity of the instincts in the higher animals are remarkable in contrast with those of the lower animals. Cuvier maintained that instinct and intelligence stand in an inverse ratio to each other; and some have thought that the intellectual faculties of the higher animals have been gradually developed from their instincts. But Pouchet, in an interesting essay,46 has shewn that no such inverse ratio really exists. Those insects which possess the most wonderful instincts are certainly the most intelligent. In the vertebrate series, the least intelligent members, namely fishes and amphibians, do not possess complex instincts; and amongst mammals the animal most remarkable for its instincts, namely the beaver, is highly intelligent, as will be admitted by every one who has read Mr. Morgan's excellent account of this animal.47

Although the first dawnings of intelligence, according to Mr. Herbert Spencer,48 have been developed through the multiplication and co-ordination of reflex actions, and although many of the simpler instincts graduate into actions of this kind and can hardly be distinguished from them, as in the case of young animals sucking, yet the more complex instincts seem to have originated independently of intelligence. I am, however, far from wishing to deny that instinctive actions may lose their fixed and untaught character, and be replaced by others performed by the aid of the free will. On the other hand, some intelligent actions—as when, birds on oceanic islands first learn to avoid man—after 38being performed during many generations, become converted into instincts and are inherited. They may then be said to be degraded in character, for they are no longer performed through reason or from experience. But the greater number of the more complex instincts appear to have been gained in a wholly different manner, through the natural selection of variations of simpler instinctive actions. Such variations appear to arise from the same unknown causes acting on the cerebral organisation, which induce slight variations or individual differences in other parts of the body; and these variations, owing to our ignorance, are often said to arise spontaneously. We can, I think, come to no other conclusion with respect to the origin of the more complex instincts, when we reflect on the marvellous instincts of sterile worker-ants and bees, which leave no offspring to inherit the effects of experience and of modified habits.

Although a high degree of intelligence is certainly compatible with the existence of complex instincts, as we see in the insects just named and in the beaver, it is not improbable that they may to a certain extent interfere with each other's development. Little is known about the functions of the brain, but we can perceive that as the intellectual powers become highly developed, the various parts of the brain must be connected by the most intricate channels of intercommunication; and as a consequence each separate part would perhaps tend to become less well fitted to answer in a definite and uniform, that is instinctive, manner to particular sensations or associations.

I have thought this digression worth giving, because we may easily underrate the mental powers of the higher animals, and especially of man, when we compare their actions founded on the memory of past events, on foresight, reason, and imagination, with39 exactly similar actions instinctively performed by the lower animals; in this latter case the capacity of performing such actions having been gained, step by step, through the variability of the mental organs and natural selection, without any conscious intelligence on the part of the animal during each successive generation. No doubt, as Mr. Wallace has argued,49 much of the intelligent work done by man is due to imitation and not to reason; but there is this great difference between his actions and many of those performed by the lower animals, namely, that man cannot, on his first trial, make, for instance, a stone hatchet or a canoe, through his power of imitation. He has to learn his work by practice; a beaver, on the other hand, can make its dam or canal, and a bird its nest, as well, or nearly as well, the first time it tries, as when old and experienced.

To return to our immediate subject: the lower animals, like man, manifestly feel pleasure and pain, happiness and misery. Happiness is never better exhibited than by young animals, such as puppies, kittens, lambs, &c., when playing together, like our own children. Even insects play together, as has been described by that excellent observer, P. Huber,50 who saw ants chasing and pretending to bite each other, like so many puppies.

The fact that the lower animals are excited by the same emotions as ourselves is so well established, that it will not be necessary to weary the reader by many details. Terror acts in the same manner on them as on us, causing the muscles to tremble, the heart to palpitate, the sphincters to be relaxed, and the hair to stand on end. Suspicion, the offspring of fear, is eminently characteristic of most wild animals. Courage 40and timidity are extremely variable qualities in the individuals of the same species, as is plainly seen in our dogs. Some dogs and horses are ill-tempered and easily turn sulky; others are good-tempered; and these qualities are certainly inherited. Every one knows how liable animals are to furious rage, and how plainly they show it. Many anecdotes, probably true, have been published on the long-delayed and artful revenge of various animals. The accurate Rengger and Brehm51 state that the American and African monkeys which they kept tame, certainly revenged themselves. The love of a dog for his master is notorious; in the agony of death he has been known to caress his master, and every one has heard of the dog suffering under vivisection, who licked the hand of the operator; this man, unless he had a heart of stone, must have felt remorse to the last hour of his life. As Whewell52 has remarked, "who that reads the touching instances of maternal affection, related so often of the women of all nations, and of the females of all animals, can doubt that the principle of action is the same in the two cases?"

We see maternal affection exhibited in the most trifling details; thus Rengger observed an American monkey (a Cebus) carefully driving away the flies which plagued her infant; and

Duvaucel saw a Hylobates washing the faces of her young ones in a stream. So intense is the grief of female monkeys for the loss of their young, that it invariably caused the death of certain kinds kept under confinement by Brehm in N. 41Africa. Orphan-monkeys were always adopted and carefully guarded by the other monkeys, both males and females. One female baboon had so capacious a heart that she not only adopted young monkeys of other species, but stole young dogs and cats, which she continually carried about. Her kindness, however, did not go so far as to share her food with her adopted offspring, at which Brehm was surprised, as his monkeys always divided everything quite fairly with their own young ones. An adopted kitten scratched the above-mentioned affectionate baboon, who certainly had a fine intellect, for she was much astonished at being scratched, and immediately examined the kitten's feet, and without more ado bit off the claws. In the Zoological Gardens, I heard from the keeper that an old baboon (C. chacma) had adopted a Rhesus monkey; but when a young drill and mandrill were placed in the cage, she seemed to perceive that these monkeys, though distinct species, were her nearer relatives, for she at once rejected the Rhesus and adopted both of them. The young Rhesus, as I saw, was greatly discontented at being thus rejected, and it would, like a naughty child, annoy and attack the young drill and mandrill whenever it could do so with safety; this conduct exciting great indignation in the old baboon. Monkeys will also, according to Brehm, defend their master when attacked by any one, as well as dogs to whom they are attached, from the attacks of other dogs. But we here trench on the subject of sympathy, to which I shall recur. Some of Brehm's monkeys took much delight in teasing, in various ingenious ways, a certain old dog whom they disliked, as well as other animals.

Most of the more complex emotions are common to the higher animals and ourselves. Every one has seen42 how jealous a dog is of his master's affection, if lavished on any other creature; and I have observed the same fact with monkeys. This shews that animals not only love, but have the desire to be loved. Animals manifestly feel emulation. They love approbation or praise; and a dog carrying a basket for his master exhibits in a high degree self-complacency or pride. There can, I think, be no doubt that a dog feels shame, as distinct from fear, and something very like modesty when begging too often for food. A great dog scorns the snarling of a little dog, and this may be called magnanimity. Several observers have stated that monkeys certainly dislike being laughed at; and they sometimes invent imaginary offences. In the Zoological Gardens I saw a baboon who always got into a furious rage when his keeper took out a letter or book and read it aloud to him; and his rage was so violent that, as I witnessed on one occasion, he bit his own leg till the blood flowed.

We will now turn to the more intellectual emotions and faculties, which are very important, as forming the basis for the development of the higher mental powers. Animals manifestly enjoy excitement and suffer from ennui, as may be seen with dogs, and, according to Rengger, with monkeys. All animals feel Wonder, and many exhibit Curiosity. They sometimes suffer from this latter quality, as when the hunter plays antics and thus attracts them; I have witnessed this with deer, and so it is with the wary chamois, and with some kinds of wild-ducks. Brehm gives a curious account of the instinctive dread which his monkeys exhibited towards snakes; but their curiosity was so great that they could not desist from occasionally satiating their horror in a most human fashion, by lifting up the lid of the box in which the snakes were kept. I was so much surprised at his account, that I took a stuffed and43 coiled-up snake into the monkey-house at the Zoological Gardens, and the excitement thus caused was one of the most curious spectacles which I ever beheld. Three species of Cercopithecus were the most alarmed; they dashed about their cages and uttered sharp signal-cries of danger, which were understood by the other monkeys. A few young monkeys and one old Anubis baboon alone took no notice of the snake. I then placed the stuffed specimen on the ground in one of the larger compartments. After a time all the monkeys collected round it in a large circle, and staring intently, presented a most ludicrous appearance. They became extremely nervous; so that when a wooden ball, with which they were familiar as a plaything, was accidently moved in the straw, under which it was partly hidden, they all instantly started away. These monkeys behaved very differently when a dead fish, a mouse, and some other new objects were placed in their cages; for though at first frightened, they soon approached, handled and examined them. I then placed a live snake in a paper bag, with the mouth loosely closed, in one of the larger compartments. One of the monkeys immediately approached, cautiously opened the bag a little, peeped in, and instantly dashed away. Then I witnessed what Brehm has described, for monkey after monkey, with head raised high and turned on one side, could not resist taking momentary peeps into the upright bag, at the dreadful object lying quiet at the bottom. It would almost appear as if monkeys had some notion of zoological affinities, for those kept by Brehm exhibited a strange, though mistaken, instinctive dread of innocent lizards and frogs. An orang, also, has been known to be much alarmed at the first sight of a turtle.53

44

The principle of Imitation is strong in man, and especially in man in a barbarous state. Desor54 has remarked that no animal voluntarily imitates an action performed by man, until in the ascending scale we come to monkeys, which are well-known to be ridiculous mockers. Animals, however, sometimes imitate each others' actions: thus two species of wolves, which had been reared by dogs, learned to bark, as does sometimes the jackal,55 but whether this can be called voluntary imitation is another question. From one account which I have read, there is reason to believe that puppies nursed by cats sometimes learn to lick their feet and thus to clean their faces: it is at least certain, as I hear from a perfectly trustworthy friend, that some dogs behave in this manner. Birds imitate the songs of their parents, and sometimes those of other birds; and parrots are notorious imitators of any sound which they often hear.

Hardly any faculty is more important for the intellectual progress of man than the power of Attention. Animals clearly manifest this power, as when a cat watches by a hole and prepares to spring on its prey. Wild animals sometimes become so absorbed when thus engaged, that they may be easily approached. Mr. Bartlett has given me a curious proof how variable this faculty is in monkeys. A man who trains monkeys to act used to purchase common kinds from the Zoological Society at the price of five pounds for each; but he offered to give double the price, if he might keep three or four of them for a few days, in order to select one. When asked how he could possibly so soon learn whether 45a particular monkey would turn out a good actor, he answered that it all depended on their power of attention. If when he was talking and explaining anything to a monkey, its attention was easily distracted, as by a fly on the wall or other trifling object, the case was hopeless. If he tried by punishment to make an inattentive monkey act, it turned sulky. On the other hand, a monkey which carefully attended to him could always be trained.

It is almost superfluous to state that animals have excellent Memories for persons and places. A baboon at the Cape of Good Hope, as I have been informed by Sir Andrew Smith, recognised him with joy after an absence of nine months. I had a dog who was savage and averse to all strangers, and I purposely tried his memory after an absence of five years and two days. I went near the stable where he lived, and shouted to him in my old manner; he showed no joy, but instantly followed me out walking and obeyed me, exactly as if I had parted with him only half-an-hour before. A train of old associations, dormant during five years, had thus been instantaneously awakened in his mind. Even ants, as P. Huber[56]has clearly shewn, recognised their fellow-ants belonging to the same community after a separation of four months. Animals can certainly by some means judge of the intervals of time between recurrent events.

The Imagination is one of the highest prerogatives of man. By this faculty he unites, independently of the will, former images and ideas, and thus creates brilliant and novel results. A poet, as Jean Paul Richter remarks,[57] "who must reflect whether he shall make a 46character say yes or no—to the devil with him; he is only a stupid corpse." Dreaming gives us the best notion of this power; as Jean Paul again says, "The dream is an involuntary art of poetry." The value of the products of our imagination depends of course on the number, accuracy, and clearness of our impressions; on our judgment and taste in selecting or rejecting the involuntary combinations, and to a certain extent on our power of voluntarily combining them. As dogs, cats, horses, and probably all the higher animals, even birds, as is stated on good authority,[58] have vivid dreams, and this is shewn by their movements and voice, we must admit that they possess some power of imagination.

Of all the faculties of the human mind, it will, I presume, be admitted that Reason stands at the summit. Few persons any longer dispute that animals possess some power of reasoning. Animals may constantly be seen to pause, deliberate, and resolve. It is a significant fact, that the more the habits of any particular animal are studied by a naturalist, the more he attributes to reason and the less to unlearnt instincts.[59] In future chapters we shall see that some animals extremely low in the scale apparently display a certain amount of reason. No doubt it is often difficult to distinguish between the power of reason and that of instinct. Thus Dr. Hayes, in his work on 'The Open Polar Sea,' repeatedly remarks that his dogs, instead of continuing to draw the sledges in a compact body, diverged and separated when they came to thin ice, so that their weight might be more evenly distributed. This was often the first warn47ing and notice which the travellers received that the ice was becoming thin and dangerous. Now, did the dogs act thus from the experience of each individual, or from the example of the older and wiser dogs, or from an inherited habit, that is from an instinct? This instinct might possibly have arisen since the time, long ago, when dogs were first employed by the natives in drawing their sledges; or the Arctic wolves, the parent-stock of the Esquimaux dog, may have acquired this instinct, impelling them not to attack their prey in a close pack when on thin ice. Questions of this kind are most difficult to answer.

So many facts have been recorded in various works shewing that animals possess some degree of reason, that I will here give only two or three instances, authenticated by Rengger, and relating to American monkeys, which stand low in their order. He states that when he first gave eggs to his monkeys, they smashed them and thus lost much of their contents; afterwards they gently hit one end against some hard body, and picked off the bits of shell with their fingers. After cutting themselves only once with any sharp tool, they would not touch it again, or would handle it with the greatest care. Lumps of sugar were often given them wrapped up in paper; and Rengger sometimes put a live wasp in the paper, so that in hastily unfolding it they got stung; after this had once happened, they always first held the packet to their ears to detect any movement within. Any one who is not convinced by such facts as these, and by what he may observe with his own dogs, that animals can reason, would not be convinced by anything that I could add. Nevertheless I will give one case with respect to dogs, as it rests on two distinct observers, and can hardly depend on the modification of any instinct.

48Mr. Colquhoun60 winged two wild-ducks, which fell on the opposite side of a stream; his retriever tried to bring over both at once, but could not succeed; she then, though never before known to ruffle a feather, deliberately killed one, brought over the other, and returned for the dead bird. Col. Hutchinson relates that two partridges were shot at once, one being killed, the other wounded; the latter ran away, and was caught by the retriever, who on her return came across the dead bird; "she stopped, evidently greatly puzzled, and after one or two trials, finding she could not take it up without permitting the escape of the winged bird, she considered a moment, then deliberately murdered it by giving it a severe crunch, and afterwards brought away both together. This was the only known instance of her ever having wilfully injured any game." Here we have reason, though not quite perfect, for the retriever might have brought the wounded bird first and then returned for the dead one, as in the case of the two wild-ducks.

The muleteers in S. America say, "I will not give you the mule whose step is easiest, but la mas rational,—the one that reasons best;" and Humboldt61 adds, "this popular expression, dictated by long experience, combats the system of animated machines, better perhaps than all the arguments of speculative philosophy."

It has, I think, now been shewn that man and the higher animals, especially the Primates, have some few instincts in common. All have the same senses, intuitions and sensations— similar passions, affections, and emotions, even the more complex ones; they feel 49wonder and curiosity; they possess the same faculties of imitation, attention, memory, imagination, and reason, though in very different degrees. Nevertheless many authors have insisted that man is separated through his mental faculties by an impassable barrier from all the lower animals. I formerly made a collection of above a score of such aphorisms, but they are not worth giving, as their wide difference and number prove the difficulty, if not the impossibility, of the attempt. It has been asserted that man alone is capable of progressive improvement; that he alone makes use of tools or fire, domesticates other animals, possesses property, or employs language; that no other animal is self-conscious, comprehends itself, has the power of abstraction, or possesses general ideas; that man alone has a sense of beauty, is liable to caprice, has the feeling of gratitude, mystery, &c.; believes in God, or is endowed with a conscience. I will hazard a few remarks on the more important and interesting of these points.

Archbishop Sumner formerly maintained62 that man alone is capable of progressive improvement. With animals, looking first to the individual, every one who has had any experience in setting traps knows that young animals can be caught much more easily than old ones; and they can be much more easily approached by an enemy. Even with respect to old animals, it is impossible to catch many in the same place and in the same kind of trap, or to destroy them by the same kind of poison; yet it is improbable that all should have partaken of the poison, and impossible that all should have been caught in the trap. They must learn caution by seeing their brethren caught or poisoned. In North America, where the fur-bearing animals have long been 50pursued, they exhibit, according to the unanimous testimony of all observers, an almost incredible amount of sagacity, caution, and cunning; but trapping has been there so long carried on that inheritance may have come into play.

If we look to successive generations, or to the race, there is no doubt that birds and other animals gradually both acquire and lose caution in relation to man or other enemies;63 and this caution is certainly in chief part an inherited habit or instinct, but in part the result of individual experience. A good observer, Leroy,[64] states that in districts where foxes are much hunted, the young when they first leave their burrows are incontestably much more wary than the old ones in districts where they are not much disturbed.

Our domestic dogs are descended from wolves and jackals,[65] and though they may not have gained in cunning, and may have lost in waryness and suspicion, yet they have progressed in certain moral qualities, such as in affection, trust-worthiness, temper, and probably in general intelligence. The common rat has conquered and beaten several other species throughout Europe, in parts of North America, New Zealand, and recently in Formosa, as well as on the mainland of China. Mr. Swinhoe,[66] who describes these latter cases, attributes the victory of the common rat over the large Mus coninga to its superior cunning; and this latter quality may be attributed to the habitual exercise of all its faculties in avoiding extirpation by man, as well 51as to nearly all the less cunning or weak-minded rats having been successively destroyed by him. To maintain, independently of any direct evidence, that no animal during the course of ages has progressed in intellect or other mental faculties, is to beg the question of the evolution of species. Hereafter we shall see that, according to Lartet, existing mammals belonging to several orders have larger brains than their ancient tertiary prototypes.

It has often been said that no animal uses any tool; but the chimpanzee in a state of nature cracks a native fruit, somewhat like a walnut, with a stone.[67] Rengger[68] easily taught an American monkey thus to break open hard palm-nuts, and afterwards of its own accord it used stones to open other kinds of nuts, as well as boxes. It thus also removed the soft rind of fruit that had a disagreeable flavour. Another monkey was taught to open the lid of a large box with a stick, and afterwards it used the stick as a lever to move heavy bodies; and I have myself seen a young orang put a stick into a crevice, slip his hand to the other end, and use it in the proper manner as a lever. In the cases just mentioned stones and sticks were employed as implements; but they are likewise used as weapons. Brehm[69] states, on the authority of the well-known traveller Schimper, that in Abyssinia when the baboons belonging to one species (C. gelada) descend in troops from the mountains to plunder the fields, they sometimes encounter troops of another species (C. hamadryas), and then a fight ensues. The Geladas roll down great stones, which the Hamadryas try to avoid, and then, both species, 52making a great uproar, rush furiously against each other. Brehm, when accompanying the Duke of Coburg-Gotha, aided in an attack with fire-arms on a troop of baboons in the pass of Mensa in Abyssinia. The baboons in return rolled so many stones down the mountain, some as large as a man's head, that the attackers had to beat a hasty retreat; and the pass was actually for a time closed against the caravan. It deserves notice that these baboons thus acted in concert. Mr. Wallace[70] on three occasions saw female orangs, accompanied by their young, "breaking off branches and the great spiny fruit of the Durian tree, with every appearance of rage; causing such a shower of missiles as effectually kept us from approaching too near the tree."

In the Zoological Gardens a monkey which had weak teeth used to break open nuts with a stone; and I was assured by the keepers that this animal, after using the stone, hid it in the straw, and would not let any other monkey touch it. Here, then, we have the idea of property; but this idea is common to every dog with a bone, and to most or all birds with their nests.

The Duke of Argyll[71] remarks, that the fashioning of an implement for a special purpose is absolutely peculiar to man; and he considers that this forms an immeasurable gulf between him and the brutes. It is no doubt a very important distinction, but there appears to me much truth in Sir J. Lubbock's suggestion,[72] that when primeval man first used flint-stones for any purpose, he would have accidentally splintered them, and would then have used the sharp fragments. From this step it would be a small one to intentionally break the 53flints, and not a very wide step to rudely fashion them. This latter advance, however, may have taken long ages, if we may judge by the immense interval of time which elapsed before the men of the neolithic period took to grinding and polishing their stone tools. In breaking the flints, as Sir J. Lubbock likewise remarks, sparks would have been emitted, and in grinding them heat would have been evolved: "thus the two usual methods of obtaining fire may have originated." The nature of fire would have been known in the many volcanic regions where lava occasionally flows through forests. The anthropomorphous apes, guided probably by instinct, build for themselves temporary platforms; but as many instincts are largely controlled by reason, the simpler ones, such as this of building a platform, might readily pass into a voluntary and conscious act. The orang is known to cover itself at night with the leaves of the Pandanus; and Brehm states that one of his baboons used to protect itself from the heat of the sun by throwing a straw-mat over its head. In these latter habits, we probably see the first steps towards some of the simpler arts; namely rude architecture and dress, as they arose amongst the early progenitors of man.

Language.—This faculty has justly been considered as one of the chief distinctions between man and the lower animals. But man, as a highly competent judge, Archbishop Whately remarks, "is not the only animal that can make use of language to express what is passing in his mind, and can understand, more or less, what is so expressed by another."[73] In Paraguay the Cebus Azaræ when excited utters at least six distinct sounds, which 54excite in other monkeys similar emotions.[74] The movements of the features and gestures of monkeys are understood by us, and they partly understand ours, as Rengger and others declare. It is a more remarkable fact that the dog, since being domesticated, has learnt to bark[75] in at least four or five distinct tones. Although barking is a new art, no doubt the wild species, the parents of the dog, expressed their feelings by cries of various kinds. With the domesticated dog we have the bark of eagerness, as in the chase; that of anger; the yelping or howling bark of despair, as when shut up; that of joy, as when starting on a walk with his master; and the very distinct one of demand or supplication, as when wishing for a door or window to be opened.

Articulate language is, however, peculiar to man; but he uses in common with the lower animals inarticulate cries to express his meaning, aided by gestures and the movements of the muscles of the face.[76] This especially holds good with the more simple and vivid feelings, which are but little connected with our higher intelligence. Our cries of pain, fear, surprise, anger, together with their appropriate actions, and the murmur of a mother to her beloved child, are more expressive than any words. It is not the mere power of articulation that distinguishes man from other animals, for as every one knows, parrots can talk; but it is his large power of connecting definite sounds with definite ideas; and this obviously depends on the development of the mental faculties.

55

As Horne Tooke, one of the founders of the noble science of philology, observes, language is an art, like brewing or baking; but writing would have been a much more appropriate simile. It certainly is not a true instinct, as every language has to be learnt. It differs, however, widely from all ordinary arts, for man has an instinctive tendency to speak, as we see in the babble of our young children; whilst no child has an instinctive tendency to brew, bake, or write. Moreover, no philologist now supposes that any language has been deliberately invented; each has been slowly and unconsciously developed by many steps. The sounds uttered by birds offer in several respects the nearest analogy to language, for all the members of the same species utter the same instinctive cries expressive of their emotions; and all the kinds that have the power of singing exert this power instinctively; but the actual song, and even the call-notes, are learnt from their parents or foster-parents. These sounds, as Daines Barrington[77] has proved, "are no more innate than language is in man." The first attempts to sing "may be compared to the imperfect endeavour in a child to babble." The young males continue practising, or, as the bird-catchers say, recording, for ten or eleven months. Their first essays show hardly a rudiment of the future song; but as they grow older we can perceive what they are aiming at; and at last they are said "to sing their song round." Nestlings which have learnt the song of a distinct species, as with the canary-birds educated in the Tyrol, teach and transmit their new song to their offspring. The slight natural differences of song in the same species inha56biting different districts may be appositely compared, as Barrington remarks, "to provincial dialects;" and the songs of allied, though distinct species may be compared with the languages of distinct races of man. I have given the foregoing details to shew that an instinctive tendency to acquire an art is not a peculiarity confined to man.

With respect to the origin of articulate language, after having read on the one side the highly interesting works of Mr. Hensleigh Wedgwood, the Rev. F. Farrar, and Prof. Schleicher,[78] and the celebrated lectures of Prof. Max Müller on the other side, I cannot doubt that language owes its origin to the imitation and modification, aided by signs and gestures, of various natural sounds, the voices of other animals, and man's own instinctive cries. When we treat of sexual selection we shall see that primeval man, or rather some early progenitor of man, probably used his voice largely, as does one of the gibbon-apes at the present day, in producing true musical cadences, that is in singing; we may conclude from a widely-spread analogy that this power would have been especially exerted during the courtship of the sexes, serving to express various emotions, as love, jealousy, triumph, and serving as a challenge to their rivals. The imitation by articulate sounds of musical cries might have given rise to words expressive of various complex emotions. As bearing on the subject of imitation, the strong tendency in our nearest allies, the monkeys, in microcephalous 57idiots,[79] and in the barbarous races of mankind, to imitate whatever they hear deserves notice. As monkeys certainly understand much that is said to them by man, and as in a state of nature they utter signal-cries of danger to their fellows,[80] it does not appear altogether incredible, that some unusually wise ape-like animal should have thought of imitating the growl of a beast of prey, so as to indicate to his fellow monkeys the nature of the expected danger. And this would have been a first step in the formation of a language. As the voice was used more and more, the vocal organs would have been strengthened and perfected through the principle of the inherited effects of use; and this would have reacted on the power of speech. But the relation between the continued use of language and the development of the brain has no doubt been far more important. The mental powers in some early progenitor of man must have been more highly developed than in any existing ape, before even the most imperfect form of speech could have come into use; but we may confidently believe that the continued use and advancement of this power would have reacted on the mind by enabling and encouraging it to carry on long trains of thought. A long and complex train of thought can no more be carried on without the aid of words, whether spoken or silent, than a long calculation without the use of figures or algebra. It appears, also, that even ordinary trains of thought almost require some form of language, for the dumb, deaf, and blind girl, Laura Bridgman, was observed to use her fingers whilst dream58ing.81 Nevertheless a long succession of vivid and connected ideas, may pass through the mind without the aid of any form of language, as we may infer from the prolonged dreams of dogs. We have, also, seen that retriever-dogs are able to reason to a certain extent; and this they manifestly do without the aid of language. The intimate connection between the brain, as it is now developed in us, and the faculty of speech, is well shewn by those curious cases of brain-disease, in which speech is specially affected, as when the power to remember substantives is lost, whilst other words can be correctly used.82 There is no more improbability in the effects of the continued use of the vocal and mental organs being inherited, than in the case of handwriting, which depends partly on the structure of the hand and partly on the disposition of the mind; and handwriting is certainly inherited.83

Why the organs now used for speech should have been originally perfected for this purpose, rather than any other organs, it is not difficult to see. Ants have considerable powers of intercommunication by means of their antennæ, as shewn by Huber, who devotes a whole chapter to their language. We might have used our fingers as efficient instruments, for a person with practice can report to a deaf man every word of a speech rapidly delivered at a public meeting; but the loss of our hands, whilst thus employed, would have been a serious inconvenience. As all the higher mammals possess vocal organs constructed on the same general 59plan with ours, and which are used as a means of communication, it was obviously probable, if the power of communication had to be improved, that these same organs would have been still further developed; and this has been effected by the aid of adjoining and well-adapted parts, namely the tongue and lips.84 The fact of the higher apes not using their vocal organs for speech, no doubt depends on their intelligence not having been sufficiently advanced. The possession by them of organs, which with long-continued practice might have been used for speech, although not thus used, is paralleled by the case of many birds which possess organs fitted for singing, though they never sing. Thus, the nightingale and crow have vocal organs similarly constructed, these being used by the former for diversified song, and by the latter merely for croaking.85

The formation of different languages and of distinct species, and the proofs that both have been developed through a gradual process, are curiously the same.86 But we can trace the origin of many words further back than in the case of species, for we can perceive that they have arisen from the imitation of various sounds, as in alliterative poetry. We find in distinct languages striking homologies due to community of descent, and analogies due to a similar process of 60formation. The manner in which certain letters or sounds change when others change is very like correlated growth. We have in both cases the reduplication of parts, the effects of long-continued use, and so forth. The frequent presence of rudiments, both in languages and in species, is still more remarkable. The letter m in the word am, means I; so that in the expression I am, a superfluous and useless rudiment has been retained. In the spelling also of words, letters often remain as the rudiments of ancient forms of pronunciation. Languages, like organic beings, can be classed in groups under groups; and they can be classed either naturally according to descent, or artificially by other characters. Dominant languages and dialects spread widely and lead to the gradual extinction of other tongues. A language, like a species, when once extinct, never, as Sir C. Lyell remarks, reappears. The same language never has two birthplaces. Distinct languages may be crossed or blended together.87 We see variability in every tongue, and new words are continually cropping up; but as there is a limit to the powers of the memory, single words, like whole languages, gradually become extinct. As Max Müller[88] has well remarked:—"A struggle for life is constantly going on amongst the words and grammatical forms in each language. The better, the shorter, the easier forms are constantly gaining the upper hand, and they owe their success to their own inherent virtue." To these more important causes of the survival of certain words, mere novelty may, I think, be added; for there is in the mind of man a strong love for slight changes in all things. The survival or[61]preservation of certain favoured words in the struggle for existence is natural selection.

The perfectly regular and wonderfully complex construction of the languages of many barbarous nations has often been advanced as a proof, either of the divine origin of these languages, or of the high art and former civilisation of their founders. Thus F. von Schlegel writes: "In those languages which appear to be at the lowest grade of intellectual culture, we frequently observe a very high and elaborate degree of art in their grammatical structure. This is especially the case with the Basque and the Lapponian, and many of the American languages."[89] But it is assuredly an error to speak of any language as an art in the sense of its having been elaborately and methodically formed. Philologists now admit that conjugations, declensions, &c., originally existed as distinct words, since joined together; and as such words express the most obvious relations between objects and persons, it is not surprising that they should have been used by the men of most races during the earliest ages. With respect to perfection, the following illustration will best shew how easily we may err: a Crinoid sometimes consists of no less than 150,000 pieces of shell,[90] all arranged with perfect symmetry in radiating lines; but a naturalist does not consider an animal of this kind as more perfect than a bilateral one with comparatively few parts, and with none of these alike, excepting on the opposite sides of the body. He justly considers the differentiation and specialisation of organs as the test of perfection. So with languages, the most symmetrical and complex ought not to be ranked above irregular, abbre[62]viated, and bastardised languages, which have borrowed expressive words and useful forms of construction from various conquering, or conquered, or immigrant races.

From these few and imperfect remarks I conclude that the extremely complex and regular construction of many barbarous languages, is no proof that they owe their origin to a special act of creation.[91] Nor, as we have seen, does the faculty of articulate speech in itself offer any insuperable objection to the belief that man has been developed from some lower form.

Self-consciousness, Individuality, Abstraction, General Ideas, &c.—It would be useless to attempt discussing these high faculties, which, according to several recent writers, make the sole and complete distinction between man and the brutes, for hardly two authors agree in their definitions. Such faculties could not have been fully developed in man until his mental powers had advanced to a high standard, and this implies the use of a perfect language. No one supposes that one of the lower animals reflects whence he comes or whither he goes,— what is death or what is life, and so forth. But can we feel sure that an old dog with an excellent memory and some power of imagination, as shewn by his dreams, never reflects on his past pleasures in the chase? and this would be a form of self-consciousness. On the other hand, as Büchner[92] has remarked, how little can the hard-worked wife of a degraded Australian savage, who uses hardly any abstract words and cannot count above four, exert her self-consciousness, or reflect on the nature of her own existence.

[63]That animals retain their mental individuality is unquestionable. When my voice awakened a train of old associations in the mind of the above-mentioned dog, he must have retained his mental individuality, although every atom of his brain had probably undergone change more than once during the interval of five years. This dog might have brought forward the

argument lately advanced to crush all evolutionists, and said, "I abide amid all mental moods and all material changes.... The teaching that atoms leave their impressions as legacies to other atoms falling into the places they have vacated is contradictory of the utterance of consciousness, and is therefore false; but it is the teaching necessitated by evolutionism, consequently the hypothesis is a false one."93

Sense of Beauty.—This sense has been declared to be peculiar to man. But when we behold male birds elaborately displaying their plumes and splendid colours before the females, whilst other birds not thus decorated make no such display, it is impossible to doubt that the females admire the beauty of their male partners. As women everywhere deck themselves with these plumes, the beauty of such ornaments cannot be disputed. The Bower-birds by tastefully ornamenting their playing-passages with gaily-coloured objects, as do certain humming-birds their nests, offer additional evidence that they possess a sense of beauty. So with the song of birds, the sweet strains poured forth by the males during the season of love are certainly admired by the females, of which fact evidence will hereafter be given. If female birds had been incapable of appreciating the beautiful colours, the ornaments, and voices 64of their male partners, all the labour and anxiety exhibited by them in displaying their charms before the females would have been thrown away; and this it is impossible to admit. Why certain bright colours and certain sounds should excite pleasure, when in harmony, cannot, I presume, be explained any more than why certain flavours and scents are agreeable; but assuredly the same colours and the same sounds are admired by us and by many of the lower animals.

The taste for the beautiful, at least as far as female beauty is concerned, is not of a special nature in the human mind; for it differs widely in the different races of man, as will hereafter be shewn, and is not quite the same even in the different nations of the same race. Judging from the hideous ornaments and the equally hideous music admired by most savages, it might be urged that their æsthetic faculty was not so highly developed as in certain animals, for instance, in birds. Obviously no animal would be capable of admiring such scenes as the heavens at night, a beautiful landscape, or refined music; but such high tastes, depending as they do on culture and complex associations, are not enjoyed by barbarians or by uneducated persons.

Many of the faculties, which have been of inestimable service to man for his progressive advancement, such as the powers of the imagination, wonder, curiosity, an undefined sense of beauty, a tendency to imitation, and the love of excitement or novelty, could not fail to have led to the most capricious changes of customs and fashions. I have alluded to this point, because a recent writer94 has oddly fixed on Caprice "as one of the most remarkable and 65typical differences between savages and brutes." But not only can we perceive how it is that roan is capricious, but the lower animals are, as we shall hereafter see, capricious in their affections, aversions, and sense of beauty. There is also good reason to suspect that they love novelty, for its own sake.

Belief in God—Religion.—There is no evidence that man was aboriginally endowed with the ennobling belief in the existence of an Omnipotent God. On the contrary there is ample evidence, derived not from hasty travellers, but from men who have long resided with savages, that numerous races have existed and still exist, who have no idea of one or more gods, and who have no words in their languages to express such an idea.95 The question is of course wholly distinct from that higher one, whether there exists a Creator and Ruler of the universe; and this has been answered in the affirmative by the highest intellects that have ever lived.

If, however, we include under the term "religion" the belief in unseen or spiritual agencies, the case is wholly different; for this belief seems to be almost universal with the less civilised races. Nor is it difficult to comprehend how it arose. As soon as the important faculties of the imagination, wonder, and curiosity, together with some power of reasoning, had become partially developed, man would naturally have craved to understand what was passing around him, and have vaguely speculated on his own existence. As 66Mr. M'Lennan96 has remarked, "Some explanation of the phenomena of life, a man must feign for himself; and to judge from the universality of it, the simplest hypothesis, and the first to occur to men, seems to have been that natural phenomena are ascribable to the presence in animals, plants, and things, and in the forces of nature, of such spirits prompting to action as men are conscious they themselves possess." It is probable, as Mr. Tylor has clearly shewn, that dreams may have first given rise to the notion of spirits; for savages do not readily distinguish between subjective and objective impressions. When a savage dreams, the figures which appear before him are believed to have come from a distance and to stand over him; or "the soul of the dreamer goes out on its travels, and comes home with a remembrance of what it has seen."97 But until the above-named faculties of imagination, curiosity, reason, &c., had been fairly well developed in the mind of man, his dreams would not have led him to believe in spirits, any more than in the case of a dog.

67

The tendency in savages to imagine that natural objects and agencies are animated by spiritual or living essences, is perhaps illustrated by a little fact which I once noticed: my dog, a full-grown and very sensible animal, was lying on the lawn during a hot and still day; but at a little distance a slight breeze occasionally moved an open parasol, which would have been wholly disregarded by the dog, had any one stood near it. As it was, every time that the parasol slightly moved, the dog growled fiercely and barked. He must, I think, have reasoned to himself in a rapid and unconscious manner, that movement without any apparent cause indicated the presence of some strange living agent, and no stranger had a right to be on his territory.

The belief in spiritual agencies would easily pass into the belief in the existence of one or more gods. For savages would naturally attribute to spirits the same passions, the same love of vengeance or simplest form of justice, and the same affections which they themselves experienced. The Fuegians appear to be in this respect in an intermediate condition, for when the surgeon on board the "Beagle" shot some young ducklings as specimens, York Minster declared in the most solemn manner, "Oh! Mr. Bynoe, much rain, much snow, blow much;" and this was evidently a retributive punishment for wasting human food. So again he related how, when his brother killed a "wild man," storms long raged, much rain and snow fell. Yet we could never discover that the Fuegians believed in what we should call a God, or practised any religious rites; and Jemmy Button, with justifiable pride, stoutly maintained that there was no devil in his land. This latter assertion is the more remarkable, as with savages the belief in bad spirits is far more common than the belief in good spirits.68

The feeling of religious devotion is a highly complex one, consisting of love, complete submission to an exalted and mysterious superior, a strong sense of dependence,98 fear, reverence, gratitude, hope for the future, and perhaps other elements. No being could experience so complex an emotion until advanced in his intellectual and moral faculties to at least a moderately high level. Nevertheless we see some distant approach to this state of mind, in the deep love of a dog for his master, associated with complete submission, some fear, and perhaps other feelings. The behaviour of a dog when returning to his master after an absence, and, as I may add, of a monkey to his beloved keeper, is widely different from that towards their fellows. In the latter case the transports of joy appear to be somewhat less, and the sense of equality is shewn in every action. Professor Braubach[99] goes so far as to maintain that a dog looks on his master as on a god.

The same high mental faculties which first led man to believe in unseen spiritual agencies, then in fetishism, polytheism, and ultimately in monotheism, would infallibly lead him, as long as his reasoning powers remained poorly developed, to various strange superstitions and customs. Many of these are terrible to think of—such as the sacrifice of human beings to a blood-loving god; the trial of innocent persons by the ordeal of poison or fire; witchcraft, &c.—yet it is well occasionally to reflect on these superstitions, for they shew us what an infinite debt of gratitude we owe to the improvement of our reason, to science, and our 69accumulated knowledge.[100] As Sir J. Lubbock has well observed, "it is not too much to say that the horrible dread of unknown evil hangs like a thick cloud over savage life, and embitters every pleasure." These miserable and indirect consequences of our highest faculties may be compared with the incidental and occasional mistakes of the instincts of the lower animals.

CHAPTER III.

Comparison of the Mental Powers of Man and the Lower Animals—continued.

The moral sense—Fundamental proposition—The qualities of social animals—Origin of sociability—Struggle between opposed instincts—Man a social animal—The more enduring social instincts conquer other less persistent instincts—The social virtues alone regarded by savages—The self-regarding virtues acquired at a later stage of development—The importance of the judgment of the members of the same community on conduct—Transmission of moral tendencies—Summary.

I fully subscribe to the judgment of those writers[101] who maintain that of all the differences between man and the lower animals, the moral sense or conscience is by far the most important. This sense, as Mackintosh[102] remarks, "has a rightful supremacy over every other principle of human action;" it is summed up in that short but imperious word ought, so full of high significance. It is the most noble of all the attributes of man, leading him without a moment's hesitation to risk his life for that of a fellow-creature; or after due deliberation, impelled simply by the deep feeling of right or duty, to sacrifice it in some great cause. Immanuel Kant exclaims, "Duty! Wondrous thought, that workest neither by fond insinuation, flattery, nor by any threat, but merely by holding up thy naked law in the soul, and so extorting for thyself always 71reverence, if not always obedience; before whom all appetites are dumb, however secretly they rebel; whence thy original?"[103]

This great question has been discussed by many writers[104] of consummate ability; and my sole excuse for touching on it is the impossibility of here passing it over, and because, as far as I know, no one has approached it exclusively from the side of natural history. The investigation possesses, also, some independent interest, as an attempt to see how far the study of the lower animals can throw light on one of the highest psychical faculties of man.

The following proposition seems to me in a high degree probable—namely, that any animal whatever, endowed with well-marked social instincts,[105] would inevitably acquire a moral sense or conscience, as soon as 72its intellectual powers had become as well developed, or nearly as well developed, as in man. For, firstly, the social instincts lead an animal to take pleasure in the society of its fellows, to feel a certain amount of sympathy with them, and to perform various services for them. The services may be of a definite and evidently instinctive nature; or there may be only a wish and readiness, as with most of the higher social animals, to aid their fellows in certain general ways. But these feelings and services are by no means extended to all the individuals of the same species, only to those of the same association. Secondly, as soon as the mental faculties had become highly developed, images of all past actions and motives would be incessantly passing through the brain of each individual; and that feeling of dissatisfaction which invariably results, as we shall hereafter see, from any unsatisfied instinct, would arise, as often as it was perceived that the enduring and always present social instinct had yielded to some other instinct, at the time stronger, but neither enduring in its nature, nor leaving behind it a very vivid impression. It is clear that many instinctive desires, such as that of hunger, are in their nature of short duration; and after being satisfied are not readily or vividly recalled. Thirdly, after the power of language had been acquired and the wishes of the members of the same community could be distinctly expressed, the common opinion how each member ought to act for the public good, would naturally become to a large extent the guide to action. But the social instincts would still give the impulse to act for the good of the community, this impulse being strengthened, directed, and sometimes even deflected by public opinion, the power of which rests, as we shall presently see, on instinctive sympathy. Lastly, habit in the individual would ultimately play a very73 important part in guiding the conduct of each member; for the social instincts and impulses, like all other instincts, would be greatly strengthened by habit, as would obedience to the wishes and judgment of the community. These several subordinate propositions must now be discussed; and some of them at considerable length.

It may be well first to premise that I do not wish to maintain that any strictly social animal, if its intellectual faculties were to become as active and as highly developed as in man, would acquire exactly the same moral sense as ours. In the same manner as various animals have some sense of beauty, though they admire widely different objects, so they might have a sense of right and wrong, though led by it to follow widely different lines of conduct. If, for instance, to take an extreme case, men were reared under precisely the same conditions as hive-bees, there can hardly be a doubt that our unmarried females would, like the worker-bees, think it a sacred duty to kill their brothers, and mothers would strive to kill their fertile daughters; and no one would think of interfering. Nevertheless the bee, or any other social animal, would in our supposed case gain, as it appears to me, some feeling of right and wrong, or a conscience. For each individual would have an inward sense of possessing certain stronger or more enduring instincts, and others less strong or enduring; so that there would often be a struggle which impulse should be followed; and satisfaction or dissatisfaction would be felt, as past impressions were compared during their incessant passage through the mind. In this case an inward monitor would tell the animal that it would have been better to have followed the one impulse rather than the other. The one course ought to have been followed: the one74 would have been right and the other wrong; but to these terms I shall have to recur.

Sociability.—Animals of many kinds are social; we find even distinct species living together, as with some American monkeys, and with the united flocks of rooks, jackdaws, and starlings. Man shows the same feeling in his strong love for the dog, which the dog returns with interest. Every one must have noticed how miserable horses, dogs, sheep, &c. are when separated from their companions; and what affection at least the two former kinds show on their reunion. It is curious to speculate on the feelings of a dog, who will rest peacefully for hours in a room with his master or any of the family, without the least notice being taken of him; but if left for a short time by himself, barks or howls dismally. We will confine our attention to the higher social animals, excluding insects, although these aid each other in many important ways. The most common service which the higher animals perform for each other, is the warning each other of danger by means of the united senses of all. Every sportsman knows, as Dr. Jaeger remarks,[106] how difficult it is to approach animals in a herd or troop. Wild horses and cattle do not, I believe, make any danger-signal; but the attitude of any one who first discovers an enemy, warns the others. Rabbits stamp loudly on the ground with their hind-feet as a signal: sheep and chamois do the same, but with their fore-feet, uttering likewise a whistle. Many birds and some mammals post sentinels, which in the case of seals are said[107] generally to be the females. The leader of a troop of monkeys acts as the sentinel, and utters cries expressive both of danger and of safety.[108] Social 75animals perform many little services for each other: horses nibble, and cows lick each other, on any spot which itches: monkeys search for each other's external parasites; and Brehm states that after a troop of the Cercopithecus griseo-viridis has rushed through a thorny brake, each monkey stretches itself on a branch, and another monkey sitting by "conscientiously" examines its fur and extracts every thorn or burr.

Animals also render more important services to each other: thus wolves and some other beasts of prey hunt in packs, and aid each other in attacking their victims. Pelicans fish in concert. The Hamadryas baboons turn over stones to find insects, &c.; and when they come to a large one, as many as can stand round, turn it over together and share the booty. Social animals mutually defend each other. The males of some ruminants come to the front when there is danger and defend the herd with their horns. I shall also in a future chapter give cases of two young wild bulls attacking an old one in concert, and of two stallions together trying to drive away a third stallion from a troop of mares. Brehm encountered in Abyssinia a great troop of baboons which were crossing a valley: some had already ascended the opposite mountain, and some were still in the valley: the latter were attacked by the dogs, but the old males immediately hurried down from the rocks, and with mouths widely opened roared so fearfully, that the dogs precipitately retreated. They were again encouraged to the attack; but by this time all the baboons had reascended the heights, excepting a young one, about six 76months old, who, loudly calling for aid, climbed on a block of rock and was surrounded. Now one of the largest males, a true hero, came down again from the mountain, slowly went to the young one, coaxed him, and triumphantly led him away—the dogs being too much astonished to make an attack. I cannot resist giving another scene which was witnessed by this same naturalist; an eagle seized a young Cercopithecus, which, by clinging to a branch, was not at once carried off; it cried loudly for assistance, upon which the other members of the troop with much uproar rushed to the rescue, surrounded the eagle, and pulled out so many feathers, that he no longer thought of his prey, but only how to escape. This eagle, as Brehm remarks, assuredly would never again attack a monkey in a troop.

It is certain that associated animals have a feeling of love for each other which is not felt by adult and non-social animals. How far in most cases they actually sympathise with each other's pains and pleasures is more doubtful, especially with respect to the latter. Mr. Buxton, however, who had excellent means of observation,[109] states that his macaws, which lived free in Norfolk, took "an extravagant interest" in a pair with a nest, and whenever the female left it, she was surrounded by a troop "screaming horrible acclamations in her honour." It is often difficult to judge whether animals have any feeling for each other's sufferings. Who can say what cows feel, when they surround and stare intently on a dying or dead companion? That animals sometimes are far from feeling any sympathy is too certain; for they will expel a wounded animal from the herd, or gore or worry it to death. This is almost the blackest fact in natural 77history, unless indeed the explanation which has been suggested is true, that their instinct or reason leads them to expel an injured companion, lest beasts of prey, including man, should be tempted to follow the troop. In this case their conduct is not much worse than that of the North American Indians who leave their feeble comrades to perish on the plains, or the Feegeans, who, when their parents get old or fall ill, bury them alive.110

Many animals, however, certainly sympathise with each other's distress or danger. This is the case even with birds; Capt. Stansbury111 found on a salt lake in Utah an old and completely blind pelican, which was very fat, and must have been long and well fed by his companions. Mr. Blyth, as he informs me, saw Indian crows feeding two or three of their companions which were blind; and I have heard of an analogous case with the domestic cock. We may, if we choose, call these actions instinctive; but such cases are much too rare for the development of any special instinct.112 I have myself seen a dog, who never passed a great friend of his, a cat which lay sick in a basket, without giving her a few licks with his tongue, the surest sign of kind feeling in a dog.

It must be called sympathy that leads a courageous dog to fly at any one who strikes his master, as he certainly will. I saw a person pretending to beat a lady who had a very timid little dog on her lap, and the trial had never before been made. The little crea78ture instantly jumped away, but after the pretended beating was over, it was really pathetic to see how perseveringly he tried to lick his mistress's face and comfort her. Brehm113 states that when a baboon in confinement was pursued to be punished, the others tried to protect him. It must have been sympathy in the cases above given which led the baboons and Cercopitheci to defend their young comrades from the dogs and the eagle. I will give only one other instance of sympathetic and heroic conduct in a little American monkey. Several years ago a keeper at the Zoological Gardens, showed me some deep and scarcely healed wounds on the nape of his neck, inflicted on him whilst kneeling on the floor by a fierce baboon. The little American monkey, who was a warm friend of this keeper, lived in the same large compartment, and was dreadfully afraid of the great baboon. Nevertheless, as soon as he saw his friend the keeper in peril, he rushed to the rescue, and by screams and bites so distracted the baboon that the man was able to escape, after running great risk, as the surgeon who attended him thought, of his life.

Besides love and sympathy, animals exhibit other qualities which in us would be called moral; and I agree with Agassiz114 that dogs possess something very like a conscience. They certainly possess some power of self-command, and this does not appear to be wholly the result of fear. As Braubach115 remarks, a dog will refrain from stealing food in the absence of his master. Dogs have long been accepted as the very type of fidelity and obedience. All animals living in a body which defend each other or attack their enemies 79in concert, must be in some degree faithful to each other; and those that follow a leader must be in some degree obedient. When the baboons in Abyssinia116 plunder a garden, they silently follow their leader; and if an imprudent young animal makes a noise, he receives a slap from the others to teach him silence and obedience; but as soon as they are sure that there is no danger, all show their joy by much clamour.

With respect to the impulse which leads certain animals to associate together, and to aid each other in many ways, we may infer that in most cases they are impelled by the same sense of satisfaction or pleasure which they experience in performing other instinctive actions; or by the same sense of dissatisfaction, as in other cases of prevented instinctive actions. We see this in innumerable instances, and it is illustrated in a striking manner by the acquired instincts of our domesticated animals; thus a young shepherd-dog delights in driving and running round a flock of sheep, but not in worrying them; a young foxhound delights in hunting a fox, whilst some other kinds of dogs as I have witnessed, utterly disregard foxes. What a strong feeling of inward satisfaction must impel a bird, so full of activity, to brood day after day over her eggs. Migratory birds are miserable if prevented from migrating, and perhaps they enjoy starting on their long flight. Some few instincts are determined solely by painful feelings, as by fear, which leads to self-preservation, or is specially directed against certain enemies. No one, I presume, can analyse the sensations of pleasure or pain. In many cases, however, it is probable that instincts are persistently followed from the 80mere force of inheritance, without the stimulus of either pleasure or pain. A young pointer, when it first scents game, apparently cannot help pointing. A squirrel in a cage who pats the nuts which it cannot eat, as if to bury them in the ground, can hardly be thought to act thus either from pleasure or pain. Hence the common assumption that men must be impelled to every action by experiencing some pleasure or pain may be erroneous. Although a habit may be blindly and implicitly followed, independently of any pleasure or pain felt at the moment, yet if it be forcibly and abruptly checked, a vague sense of dissatisfaction is generally experienced; and this is especially true in regard to persons of feeble intellect.

It has often been assumed that animals were in the first place rendered social, and that they feel as a consequence uncomfortable when separated from each other, and comfortable whilst together; but it is a more probable view that these sensations were first developed, in order that those animals which would profit by living in society, should be induced to live together. In the same manner as the sense of hunger and the pleasure of eating were, no doubt, first acquired in order to induce animals to eat. The feeling of pleasure from society is probably an extension of the parental or filial affections; and this extension may be in chief part attributed to natural selection, but perhaps in part to mere habit. For with those animals which were benefited by living in close association, the individuals which took the greatest pleasure in society would best escape various dangers; whilst those that cared least for their comrades and lived solitary would perish in greater numbers. With respect to the origin of the parental and filial affections, which apparently lie at the basis of the social affections, it is hopeless to speculate; but we81 may infer that they have been to a large extent gained through natural selection. So it has almost certainly been with the unusual and opposite feeling of hatred between the nearest relations, as with the worker-bees which kill their brother-drones, and with the queen-bees which kill their daughter-queens; the desire to destroy, instead of loving, their nearest relations having been here of service to the community.

The all-important emotion of sympathy is distinct from that of love. A mother may passionately love her sleeping and passive infant, but she can then hardly be said to feel sympathy for it. The love of a man for his dog is distinct from sympathy, and so is that of a dog for his master. Adam Smith formerly argued, as has Mr. Bain recently, that the basis of sympathy lies in our strong retentiveness of former states of pain or pleasure. Hence, "the sight of another person enduring hunger, cold, fatigue, revives in us some recollection of these states, which are painful even in idea." We are thus impelled to relieve the sufferings of another, in order that our own painful feelings may be at the same time relieved. In like manner we are led to participate in the pleasures of others.[117] But I cannot see how this view explains the fact that sympathy is excited in an immeasurably stronger degree by a beloved than by an indifferent person. The mere 82sight of suffering, independently of love, would suffice to call up in us vivid recollections and associations. Sympathy may at first have originated in the manner above suggested; but it seems now to have become an instinct, which is especially directed towards beloved objects, in the same manner as fear with animals is especially directed against certain enemies. As sympathy is thus directed, the mutual love of the members of the same community will extend its limits. No doubt a tiger or lion feels sympathy for the sufferings of its own young, but not for any other animal. With strictly social animals the feeling will be more or less extended to all the associated members, as we know to be the case. With mankind selfishness, experience, and imitation probably add, as Mr. Bain has shewn, to the power of sympathy; for we are led by the hope of receiving good in return to perform acts of sympathetic kindness to others; and there can be no doubt that the feeling of sympathy is much strengthened by habit. In however complex a manner this feeling may have originated, as it is one of high importance to all those animals which aid and defend each other, it will have been increased, through natural selection; for those communities, which included the greatest number of the most sympathetic members, would flourish best and rear the greatest number of offspring.

In many cases it is impossible to decide whether certain social instincts have been acquired through natural selection, or are the indirect result of other instincts and faculties, such as sympathy, reason, experience, and a tendency to imitation; or again, whether they are simply the result of long-continued habit. So remarkable an instinct as the placing sentinels to warn the community of danger, can hardly have been83 the indirect result of any other faculty; it must therefore have been directly acquired. On the other hand, the habit followed by the males of some social animals, of defending the community and of attacking their enemies or their prey in concert, may perhaps have originated from mutual sympathy; but courage, and in most cases strength, must have been previously acquired, probably through natural selection.

Of the various instincts and habits, some are much stronger than others, that is, some either give more pleasure in their performance and more distress in their prevention than others; or, which is probably quite as important, they are more persistently followed through inheritance without exciting any special feeling of pleasure or pain. We are ourselves conscious that some habits are much more difficult to cure or change than others. Hence a struggle may often be observed in animals between different instincts, or between an instinct and some habitual disposition; as when a dog rushes after a hare, is rebuked, pauses, hesitates, pursues again or returns ashamed to his master; or as between the love of a female dog for her young puppies and for her master, for she may be seen to slink away to them, as if half ashamed of not accompanying her master. But the most curious instance known to me of one instinct conquering another, is the migratory instinct conquering the maternal instinct. The former is wonderfully strong; a confined bird will at the proper season beat her breast against the wires of her cage, until it is bare and bloody. It causes young salmon to leap out of the fresh water, where they could still continue to live, and thus unintentionally to commit suicide. Every one knows how strong the maternal instinct is, leading even timid birds to face great danger, though with hesitation and in opposition to the instinct of self84preservation. Nevertheless the migratory instinct is so powerful that late in the autumn swallows and house-martins frequently desert their tender young, leaving them to perish miserably in their nests.118

We can perceive that an instinctive impulse, if it be in any way more beneficial to a species than some other or opposed instinct, would be rendered the more potent of the two through natural selection; for the individuals which had it most strongly developed would survive in larger numbers. Whether this is the case with the migratory in comparison with the maternal instinct, may well be doubted. The great persistence or steady action of the former at certain seasons of the year during the whole day, may give it for a time paramount force.

Man a social animal.—Most persons admit that man is a social being. We see this in his dislike of solitude, and in his wish for society beyond that of his own family. Solitary confinement is one of the severest punishments which can be inflicted. Some authors suppose that man primevally lived in single families; but at the present day, though single families, or only two or three together, roam the solitudes of some savage lands, they are always, as far as I can discover, friendly with other families inhabiting the same district. Such families occasionally meet in council, and they unite 85for their common defence. It is no argument against savage man being a social animal, that the tribes inhabiting adjacent districts are almost always at war with each other; for the social instincts never extend to all the individuals of the same species. Judging from the analogy of the greater number of the Quadrumana, it is probable that the early ape-like progenitors of man were likewise social; but this is not of much importance for us. Although man, as he now exists, has few special instincts, having lost any which his early progenitors may have possessed, this is no reason why he should not have retained from an extremely remote period some degree of instinctive love and sympathy for his fellows. We are indeed all conscious that we do possess such sympathetic feelings;119 but our consciousness does not tell us whether they are instinctive, having originated long ago in the same manner as with the lower animals, or whether they have been acquired by each of us during our early years. As man is a social animal, it is also probable that he would inherit a tendency to be faithful to his comrades, for this quality is common to most social animals. He would in like manner possess some capacity for self-command, and perhaps of obedience to the leader of the community. He would from an inherited tendency still be willing to defend, in concert with others, his fellow-men, and would be ready to aid them in any way which did not too greatly interfere with his own welfare or his own strong desires.

The social animals which stand at the bottom of the 86scale are guided almost exclusively, and those which stand higher in the scale are largely guided, in the aid which they give to the members of the same community, by special instincts; but they are likewise in part impelled by mutual love and sympathy, assisted apparently by some amount of reason. Although man, as just remarked, has no special instincts to tell him how to aid his fellow-men, he still has the impulse, and with his improved intellectual faculties would naturally be much guided in this respect by reason and experience. Instinctive sympathy would, also, cause him to value highly the approbation of his fellow-men; for, as Mr. Bain has clearly shewn,120the love of praise and the strong feeling of glory, and the still stronger horror of scorn and infamy, "are due to the workings of sympathy." Consequently man would be greatly influenced by the wishes, approbation, and blame of his fellow-men, as expressed by their gestures and language. Thus the social instincts, which must have been acquired by man in a very rude state, and probably even by his early ape-like progenitors, still give the impulse to many of his best actions; but his actions are largely determined by the expressed wishes and judgment of his fellow-men, and unfortunately still oftener by his own strong, selfish desires. But as the feelings of love and sympathy and the power of self-command become strengthened by habit, and as the power of reasoning becomes clearer so that man can appreciate the justice of the judgments of his fellow-men, he will feel himself impelled, independently of any pleasure or pain felt at the moment, to certain lines of conduct. He may then say, I am the supreme judge of my own conduct, and in the words of Kant, I will not in my own person violate the dignity of humanity.

87The more enduring Social Instincts conquer the less Persistent Instincts.—We have, however, not as yet considered the main point, on which the whole question of the moral sense hinges. Why should a man feel that he ought to obey one instinctive desire rather than another? Why does he bitterly regret if he has yielded to the strong sense of self-preservation, and has not risked his life to save that of a fellow-creature; or why does he regret having stolen food from severe hunger?

It is evident in the first place, that with mankind the instinctive impulses have different degrees of strength; a young and timid mother urged by the maternal instinct will, without a moment's hesitation, run the greatest danger for her infant, but not for a mere fellow-creature. Many a man, or even boy, who never before risked his life for another, but in whom courage and sympathy were well developed, has, disregarding the instinct of self-preservation, instantaneously plunged into a torrent to save a drowning fellow-creature. In this case man is impelled by the same instinctive motive, which caused the heroic little American monkey, formerly described, to attack the great and dreaded baboon, to save his keeper. Such actions as the above appear to be the simple result of the greater strength of the social or maternal instincts than of any other instinct or motive; for they are performed too instantaneously for reflection, or for the sensation of pleasure or pain; though if prevented distress would be caused.

I am aware that some persons maintain that actions performed impulsively, as in the above cases, do not come under the dominion of the moral sense, and cannot be called moral. They confine this term to actions done deliberately, after a victory over opposing desires, or to actions prompted by some lofty motive. But it appears scarcely possible to draw any clear line88 of distinction of this kind; though the distinction may be real. As far as exalted motives are concerned, many instances have been recorded of barbarians, destitute of any feeling of general benevolence towards mankind, and not guided by any religious motive, who have deliberately as prisoners sacrificed their lives,121 rather than betray their comrades; and surely their conduct ought to be considered as moral. As far as deliberation and the victory over opposing motives are concerned, animals may be seen doubting between opposed instincts, as in rescuing their offspring or comrades from danger; yet their actions, though done for the good of others, are not called moral. Moreover, an action repeatedly performed by us, will at last be done without deliberation or hesitation, and can then hardly be distinguished from an instinct; yet surely no one will pretend that an action thus done ceases to be moral. On the contrary, we all feel that an act cannot be considered as perfect, or as performed in the most noble manner, unless it be done impulsively, without deliberation or effort, in the same manner as by a man in whom the requisite qualities are innate. He who is forced to overcome his fear or want of sympathy before he acts, deserves, however, in one way higher credit than the man whose innate disposition leads him to a good act without effort. As we cannot distinguish between motives, we rank all actions of a certain class as moral, when they are performed by a moral being. A moral being is one who is capable of comparing his past and future actions or motives, and of approving or disapproving of them. We have no reason to suppose that any of the lower animals have 89this capacity; therefore when a monkey faces danger to rescue its comrade, or takes charge of an orphan-monkey, we do not call its conduct moral. But in the case of man, who alone can with certainty be ranked as a moral being, actions of a certain class are called moral, whether performed deliberately after a struggle with opposing motives, or from the effects of slowly-gained habit, or impulsively through instinct.

But to return to our more immediate subject; although some instincts are more powerful than others, thus leading to corresponding actions, yet it cannot be maintained that the social instincts are ordinarily stronger in man, or have become stronger through long-continued habit, than the instincts, for instance, of self-preservation, hunger, lust, vengeance, &c. Why then does man regret, even though he may endeavour to banish any such regret, that he has followed the one natural impulse, rather than the other; and why does he further feel that he ought to regret his conduct? Man in this respect differs profoundly from the lower animals. Nevertheless we can, I think, see with some degree of clearness the reason of this difference.

Man, from the activity of his mental faculties, cannot avoid reflection: past impressions and images are incessantly passing through his mind with distinctness. Now with those animals which live permanently in a body, the social instincts are ever present and persistent. Such animals are always ready to utter the danger-signal, to defend the community, and to give aid to their fellows in accordance with their habits; they feel at all times, without the stimulus of any special passion or desire, some degree of love and sympathy for them; they are unhappy if long separated from them, and always happy to be in their company. So it is with ourselves. A man who possessed no trace90 of such feelings would be an unnatural monster. On the other hand, the desire to satisfy hunger, or any passion, such as vengeance, is in its nature temporary, and can for a time be fully satisfied. Nor is it easy, perhaps hardly possible, to call up with complete vividness the feeling, for instance, of hunger; nor indeed, as has often been remarked, of any suffering. The instinct of self-preservation is not felt except in the presence of danger; and many a coward has thought himself brave until he has met his enemy face to face. The wish for another man's property is perhaps as persistent a desire as any that can be named; but even in this case the satisfaction of actual possession is generally a weaker feeling than the desire: many a thief, if not an habitual one, after success has wondered why he stole some article.

Thus, as man cannot prevent old impressions continually repassing through his mind, he will be compelled to compare the weaker impressions of, for instance, past hunger, or of vengeance satisfied or danger avoided at the cost of other men, with the instinct of sympathy and good-will to his fellows, which is still present and ever in some degree active in his mind. He will then feel in his imagination that a stronger instinct has yielded to one which now seems comparatively weak; and then that sense of dissatisfaction will inevitably be felt with which man is endowed, like every other animal, in order that his instincts may be obeyed. The case before given, of the swallow, affords an illustration, though of a reversed nature, of a temporary though for the time strongly persistent instinct conquering another instinct which is usually dominant over all others. At the proper season these birds seem all day long to be impressed with the desire to migrate; their habits change; they become restless, are noisy,91 and congregate in flocks. Whilst the mother-bird is feeding or brooding over her nestlings, the maternal instinct is probably stronger than the migratory; but the instinct which is more persistent gains the victory, and at last, at a moment when her young ones are not in sight, she takes flight and deserts them. When arrived at the end of her long journey, and the migratory instinct ceases to act, what an agony of remorse each bird would feel, if, from being endowed with great mental activity, she could not prevent the image continually passing before her mind of her young ones perishing in the bleak north from cold and hunger.

At the moment of action, man will no doubt be apt to follow the stronger impulse; and though this may occasionally prompt him to the noblest deeds, it will far more commonly lead him to gratify his own desires at the expense of other men. But after their gratification, when past and weaker impressions are contrasted with the ever-enduring social instincts, retribution will surely come. Man will then feel dissatisfied with himself, and will resolve with more or less force to act differently for the future. This is conscience; for conscience looks backwards and judges past actions, inducing that kind of dissatisfaction, which if weak we call regret, and if severe remorse.

These sensations are, no doubt, different from those experienced when other instincts or desires are left unsatisfied; but every unsatisfied instinct has its own proper prompting sensation, as we recognise with hunger, thirst, &c. Man thus prompted, will through long habit acquire such perfect self-command, that his desires and passions will at last instantly yield to his social sympathies, and there will no longer be a struggle between them. The still hungry, or the still revengeful man will not think of stealing food, or of wreaking his92vengeance. It is possible, or, as we shall hereafter see, even probable, that the habit of self-command may, like other habits, be inherited. Thus at last man comes to feel, through acquired and perhaps inherited habit, that it is best for him to obey his more persistent instincts. The imperious word ought seems merely to imply the consciousness of the existence of a persistent instinct, either innate or partly acquired, serving him as a guide, though liable to be disobeyed. We hardly use the word ought in a metaphorical sense, when we say hounds ought to hunt, pointers to point, and retrievers to retrieve their game. If they fail thus to act, they fail in their duty and act wrongly.

If any desire or instinct, leading to an action opposed to the good of others, still appears to a man, when recalled to mind, as strong as, or stronger than, his social instinct, he will feel no keen regret at having followed it; but he will be conscious that if his conduct were known to his fellows, it would meet with their disapprobation; and few are so destitute of sympathy as not to feel discomfort when this is realised. If he has no such sympathy, and if his desires leading to bad actions are at the time strong, and when recalled are not overmastered by the persistent social instincts, then he is essentially a bad man;122 and the sole restraining motive left is the fear of punishment, and the conviction that in the long run it would be best for his own selfish interests to regard the good of others rather than his own.

It is obvious that every one may with an easy conscience gratify his own desires, if they do not interfere 93with his social instincts, that is with the good of others; but in order to be quite free from self-reproach, or at least of anxiety, it is almost necessary for him to avoid the disapprobation, whether reasonable or not, of his fellow men. Nor must he break through the fixed habits of his life, especially if these are supported by reason; for if he does, he will assuredly feel dissatisfaction. He must likewise avoid the reprobation of the one God or gods, in whom according to his knowledge or superstition he may believe; but in this case the additional fear of divine punishment often supervenes.

The strictly Social Virtues at first alone regarded.—The above view of the first origin and nature of the moral sense, which tells us what we ought to do, and of the conscience which reproves us if we disobey it, accords well with what we see of the early and undeveloped condition of this faculty in mankind. The virtues which must be practised, at least generally, by rude men, so that they may associate in a body, are those which are still recognised as the most important. But they are practised almost exclusively in relation to the men of the same tribe; and their opposites are not regarded as crimes in relation to the men of other tribes. No tribe could hold together if murder, robbery, treachery, &c., were common; consequently such crimes within the limits of the same tribe "are branded with everlasting infamy;"123 but excite no such sentiment beyond these limits. A North-American Indian is well pleased with himself, and is honoured by others, when he scalps a man of another tribe; and a Dyak 94cuts off the head of an unoffending person and dries it as a trophy. The murder of infants has prevailed on the largest scale throughout the world,124 and has met with no reproach; but infanticide, especially of females, has been thought to be good for the tribe, or at least not injurious. Suicide during former times was not generally considered as a crime,125 but rather from the courage displayed as an honourable act; and it is still largely practised by some semi-civilised nations without reproach, for the loss to a nation of a single individual is not felt: whatever the explanation may be, suicide, as I hear from Sir J. Lubbock, is rarely practised by the lowest barbarians. It has been recorded that an Indian Thug conscientiously regretted that he had not strangled and robbed as many travellers as did his father before him. In a rude state of civilisation the robbery of strangers is, indeed, generally considered as honourable.

The great sin of Slavery has been almost universal, and slaves have often been treated in an infamous manner. As barbarians do not regard the opinion of their women, wives are commonly treated like slaves. Most savages are utterly indifferent to the sufferings of strangers, or even delight in witnessing them. It is well known that the women and children of the North-American Indians aided in torturing their enemies. Some savages take a horrid pleasure in cruelty to animals,126 and humanity with them is an unknown virtue. Nevertheless, feelings of sympathy and kindness are common, especially 95during sickness, between the members of the same tribe, and are sometimes extended beyond the limits of the tribe. Mungo Park's touching account of the kindness of the negro women of the interior to him is well known. Many instances could be given of the noble fidelity of savages towards each other, but not to strangers; common experience justifies the maxim of the Spaniard, "Never, never trust an Indian." There cannot be fidelity without truth; and this fundamental virtue is not rare between the members of the same tribe: thus Mungo Park heard the negro women teaching their young children to love the truth. This, again, is one of the virtues which becomes so deeply rooted in the mind that it is sometimes practised by savages even at a high cost, towards strangers; but to lie to your enemy has rarely been thought a sin, as the history of modern diplomacy too plainly shews. As soon as a tribe has a recognised leader, disobedience becomes a crime, and even abject submission is looked at as a sacred virtue.

As during rude times no man can be useful or faithful to his tribe without courage, this quality has universally been placed in the highest rank; and although, in civilised countries, a good, yet timid, man may be far more useful to the community than a brave one, we cannot help instinctively honouring the latter above a coward, however benevolent. Prudence, on the other hand, which does not concern the welfare of others, though a very useful virtue, has never been highly esteemed. As no man can practise the virtues necessary for the welfare of his tribe without self-sacrifice, self-command, and the power of endurance, these qualities have been at all times highly and most justly valued. The American savage voluntarily submits without a groan to the most horrid tortures to prove and strengthen his fortitude and courage; and we cannot96 help admiring him, or even an Indian Fakir, who, from a foolish religious motive, swings suspended by a hook buried in his flesh.

The other self-regarding virtues, which do not obviously, though they may really, affect the welfare of the tribe, have never been esteemed by savages, though now highly appreciated by civilised nations. The greatest intemperance with savages is no reproach. Their utter licentiousness, not to mention unnatural crimes, is something astounding.127 As soon, however, as marriage, whether polygamous or monogamous, becomes common, jealousy will lead to the inculcation of female virtue; and this being honoured will tend to spread to the unmarried females. How slowly it spreads to the male sex we see at the present day. Chastity eminently requires self-command; therefore it has been honoured from a very early period in the moral history of civilised man. As a consequence of this, the senseless practice of celibacy has been ranked from a remote period as a virtue.128 The hatred of indecency, which appears to us so natural as to be thought innate, and which is so valuable an aid to chastity, is a modern virtue, appertaining exclusively, as Sir G. Staunton remarks,129 to civilised life. This is shewn by the ancient religious rites of various nations, by the drawings on the walls of Pompeii, and by the practices of many savages.

We have now seen that actions are regarded by savages, and were probably so regarded by primeval man, as good or bad, solely as they affect in an obvious manner the welfare of the tribe,—not that of the species, nor that of man as an individual member of the 97tribe. This conclusion agrees well with the belief that the so-called moral sense is aboriginally derived from the social instincts, for both relate at first exclusively to the community. The chief causes of the low morality of savages, as judged by our standard, are, firstly, the confinement of sympathy to the same tribe. Secondly, insufficient powers of reasoning, so that the bearing of many virtues, especially of the self-regarding virtues, on the general welfare of the tribe is not recognised. Savages, for instance, fail to trace the multiplied evils consequent on a want of temperance, chastity, &c. And, thirdly, weak power of self-command; for this power has not been strengthened through long-continued, perhaps inherited, habit, instruction and religion.

I have entered into the above details on the immorality of savages,130 because some authors have recently taken a high view of their moral nature, or have attributed most of their crimes to mistaken benevolence.131 These authors appear to rest their conclusion on savages possessing, as they undoubtedly do possess, and often in a high degree, those virtues which are serviceable, or even necessary, for the existence of a tribal community.

Concluding Remarks.—Philosophers of the derivative132 school of morals formerly assumed that the foundation of morality lay in a form of Selfishness; but more recently in the "Greatest Happiness principle." According to the view given above, the moral sense is 98fundamentally identical with the social instincts; and in the case of the lower animals it would be absurd to speak of these instincts as having been developed from selfishness, or for the happiness of the community. They have, however, certainly been developed for the general good of the community. The term, general good, may be defined as the means by which the greatest possible number of individuals can be reared in full vigour and health, with all their faculties perfect, under the conditions to which they are exposed. As the social instincts both of man and the lower animals have no doubt been developed by the same steps, it would be advisable, if found practicable, to use the same definition in both cases, and to take as the test of morality, the general good or welfare of the community, rather than the general happiness; but this definition would perhaps require some limitation on account of political ethics.

When a man risks his life to save that of a fellow-creature, it seems more appropriate to say that he acts for the general good or welfare, rather than for the general happiness of mankind. No doubt the welfare and the happiness of the individual usually coincide; and a contented, happy tribe will flourish better than one that is discontented and unhappy. We have seen that at an early period in the history of man, the expressed wishes of the community will have naturally influenced to a large extent the conduct of each member; and as all wish for happiness, the "greatest happiness principle" will have become a most important secondary guide and object; the social instincts, including sympathy, always serving as the primary impulse and guide. Thus the reproach of laying the foundation of the most noble part of our nature in the base principle of selfishness is removed; unless indeed the satis99faction which every animal feels when it follows its proper instincts, and the dissatisfaction felt when prevented, be called selfish.

The expression of the wishes and judgment of the members of the same community, at first by oral and afterwards by written language, serves, as just remarked, as a most important secondary guide of conduct, in aid of the social instincts, but sometimes in opposition to them. This latter fact is well exemplified by the Law of Honour, that is the law of the opinion of our equals, and not of all our countrymen. The breach of this law, even when the breach is known to be strictly accordant with true morality, has caused many a man more agony than a real crime. We recognise the same influence in the burning sense of shame which most of us have felt even after the interval of years, when calling to mind some accidental breach of a trifling though fixed rule of etiquette. The judgment of the community will generally be guided by some rude experience of what is best in the long run for all the members; but this judgment will not rarely err from ignorance and from weak powers of reasoning. Hence the strangest customs and superstitions, in complete opposition to the true welfare and happiness of mankind, have become all-powerful throughout the world. We see this in the horror felt by a Hindoo who breaks his caste, in the shame of a Mahometan woman who exposes her face, and in innumerable other instances. It would be difficult to distinguish between the remorse felt by a Hindoo who has eaten unclean food, from that felt after committing a theft; but the former would probably be the more severe.

How so many absurd rules of conduct, as well as so many absurd religious beliefs, have originated we do not know; nor how it is that they have become, in all100 quarters of the world, so deeply impressed on the mind of men; but it is worthy of remark that a belief constantly inculcated during the early years of life, whilst the brain is impressible, appears to acquire almost the nature of an instinct; and the very essence of an instinct is that it is followed independently of reason. Neither can we say why certain admirable virtues, such as the love of truth, are much more highly appreciated by some savage tribes than by others;133 nor, again, why similar differences prevail even amongst civilised nations. Knowing how firmly fixed many strange customs and superstitions have become, we need feel no surprise that the self-regarding virtues should now appear to us so natural, supported as they are by reason, as to be thought innate, although they were not valued by man in his early condition.

Notwithstanding many sources of doubt, man can generally and readily distinguish between the higher and lower moral rules. The higher are founded on the social instincts, and relate to the welfare of others. They are supported by the approbation of our fellow-men and by reason. The lower rules, though some of them when implying self-sacrifice hardly deserve to be called lower, relate chiefly to self, and owe their origin to public opinion, when matured by experience and cultivated; for they are not practised by rude tribes.

As man advances in civilisation, and small tribes are united into larger communities, the simplest reason would tell each individual that he ought to extend his social instincts and sympathies to all the members of the same nation, though personally unknown to him. This point being once reached, there is only an arti101ficial barrier to prevent his sympathies extending to the men of all nations and races. If, indeed, such men are separated from him by great differences in appearance or habits, experience unfortunately shews us how long it is before we look at them as our fellow-creatures. Sympathy beyond the confines of man, that is humanity to the lower animals, seems to be one of the latest moral acquisitions. It is apparently unfelt by savages, except towards their pets. How little the old Romans knew of it is shewn by their abhorrent gladiatorial exhibitions. The very idea of humanity, as far as I could observe, was new to most of the Gauchos of the Pampas. This virtue, one of the noblest with which man is endowed, seems to arise incidentally from our sympathies becoming more tender and more widely diffused, until they are extended to all sentient beings. As soon as this virtue is honoured and practised by some few men, it spreads through instruction and example to the young, and eventually through public opinion.

The highest stage in moral culture at which we can arrive, is when we recognise that we ought to control our thoughts, and "not even in inmost thought to think again the sins that made the past so pleasant to us."134 Whatever makes any bad action familiar to the mind, renders its performance by so much the easier. As Marcus Aurelius long ago said, "Such as are thy habitual thoughts, such also will be the character of thy mind; for the soul is dyed by the thoughts."135

Our great philosopher, Herbert Spencer, has recently explained his views on the moral sense. He says,136 "I 102believe that the experiences of utility organised and consolidated through all past generations of the human race, have been producing corresponding modifications, which, by continued transmission and accumulation, have become in us certain faculties of moral intuition—certain emotions responding to right and wrong conduct, which have no apparent basis in the individual experiences of utility." There is not the least inherent improbability, as it seems to me, in virtuous tendencies being more or less strongly inherited; for, not to mention the various dispositions and habits transmitted by many of our domestic animals, I have heard of cases in which a desire to steal and a tendency to lie appeared to run in families of the upper ranks; and as stealing is so rare a crime in the wealthy classes, we can hardly account by accidental coincidence for the tendency occurring in two or three members of the same family. If bad tendencies are transmitted, it is probable that good ones are likewise transmitted. Excepting through the principle of the transmission of moral tendencies, we cannot understand the differences believed to exist in this respect between the various races of mankind. We have, however, as yet, hardly sufficient evidence on this head.

Even the partial transmission of virtuous tendencies would be an immense assistance to the primary impulse derived directly from the social instincts, and indirectly from the approbation of our fellow-men. Admitting for the moment that virtuous tendencies are inherited, it appears probable, at least in such cases as chastity, temperance, humanity to animals, &c., that they become first impressed on the mental organisation through habit, instruction, and example, continued during several generations in the same family, and in a quite subordinate degree, or not at all, by the individuals pos103sessing such virtues, having succeeded best in the struggle for life. My chief source of doubt with respect to any such inheritance, is that senseless customs, superstitions, and tastes, such as the horror of a Hindoo for unclean food, ought on the same principle to be transmitted. Although this in itself is perhaps not less probable than that animals should acquire inherited tastes for certain kinds of food or fear of certain foes, I have not met with any evidence in support of the transmission of superstitious customs or senseless habits.

Finally, the social instincts which no doubt were acquired by man, as by the lower animals, for the good of the community, will from the first have given to him some wish to aid his fellows, and some feeling of sympathy. Such impulses will have served him at a very early period as a rude rule of right and wrong. But as man gradually advanced in intellectual power and was enabled to trace the more remote consequences of his actions; as he acquired sufficient knowledge to reject baneful customs and superstitions; as he regarded more and more not only the welfare but the happiness of his fellow-men; as from habit, following on beneficial experience, instruction, and example, his sympathies became more tender and widely diffused, so as to extend to the men of all races, to the imbecile, the maimed, and other useless members of society, and finally to the lower animals,—so would the standard of his morality rise higher and higher. And it is admitted by moralists of the derivative school and by some intuitionists, that the standard of morality has risen since an early period in the history of man.137

As a struggle may sometimes be seen going on 104between the various instincts of the lower animals, it is not surprising that there should be a struggle in man between his social instincts, with their derived virtues, and his lower, though at the moment, stronger impulses or desires. This, as Mr. Galton138 has remarked, is all the less surprising, as man has emerged from a state of barbarism within a comparatively recent period. After having yielded to some temptation we feel a sense of dissatisfaction, analogous to that felt from other unsatisfied instincts, called in this case conscience; for we cannot prevent past images and impressions continually passing through our minds, and these in their weakened state we compare with the ever-present social instincts, or with habits gained in early youth and strengthened during our whole lives, perhaps inherited, so that they are at last rendered almost as strong as instincts. Looking to future generations, there is no cause to fear that the social instincts will grow weaker, and we may expect that virtuous habits will grow stronger, becoming perhaps fixed by inheritance. In this case the struggle between our higher and lower impulses will be less severe, and virtue will be triumphant.

Summary of the two last Chapters.—There can be no doubt that the difference between the mind of the lowest man and that of the highest animal is immense. An anthropomorphous ape, if he could take a dispassionate view of his own case, would admit that though he could form an artful plan to plunder a garden—though he could use stones for fighting or for breaking open nuts, 105yet that the thought of fashioning a stone into a tool was quite beyond his scope. Still less, as he would admit, could he follow out a train of metaphysical reasoning, or solve a mathematical problem, or reflect on God, or admire a grand natural scene. Some apes, however, would probably declare that they could and did admire the beauty of the coloured skin and fur of their partners in marriage. They would admit, that though they could make other apes understand by cries some of their perceptions and simpler wants, the notion of expressing definite ideas by definite sounds had never crossed their minds. They might insist that they were ready to aid their fellow-apes of the same troop in many ways, to risk their lives for them, and to take charge of their orphans; but they would be forced to acknowledge that disinterested love for all living creatures, the most noble attribute of man, was quite beyond their comprehension.

Nevertheless the difference in mind between man and the higher animals, great as it is, is certainly one of degree and not of kind. We have seen that the senses and intuitions, the various emotions and faculties, such as love, memory, attention, curiosity, imitation, reason,

&c., of which man boasts, may be found in an incipient, or even sometimes in a well-developed condition, in the lower animals. They are also capable of some inherited improvement, as we see in the domestic dog compared with the wolf or jackal. If it be maintained that certain powers, such as self-consciousness, abstraction, &c., are peculiar to man, it may well be that these are the incidental results of other highly-advanced intellectual faculties; and these again are mainly the result of the continued use of a highly developed language. At what age does the new-born infant possess the power of abstraction, or become self106conscious and reflect on its own existence? We cannot answer; nor can we answer in regard to the ascending organic scale. The half-art and half-instinct of language still bears the stamp of its gradual evolution. The ennobling belief in God is not universal with man; and the belief in active spiritual agencies naturally follows from his other mental powers. The moral sense perhaps affords the best and highest distinction between man and the lower animals; but I need not say anything on this head, as I have so lately endeavoured to shew that the social instincts,—the prime principle of man's moral constitution139—with the aid of active intellectual powers and the effects of habit, naturally lead to the golden rule, "As ye would that men should do to you, do ye to them likewise;" and this lies at the foundation of morality.

In a future chapter I shall make some few remarks on the probable steps and means by which the several mental and moral faculties of man have been gradually evolved. That this at least is possible ought not to be denied, when we daily see their development in every infant; and when we may trace a perfect gradation from the mind of an utter idiot, lower than that of the lowest animal, to the mind of a Newton.

CHAPTER IV.
On the Manner of Development of Man from some lower Form.

Variability of body and mind in man—Inheritance—Causes of variability—Laws of variation the same in man as in the lower animals—Direct action of the conditions of life—Effects of the increased use and disuse of parts—Arrested development—Reversion—Correlated variation—Rate of increase—Checks to increase—Natural selection—Man the most dominant animal in the world—Importance of his corporeal structure—The causes which have led to his becoming erect—Consequent changes of structure—Decrease in size of the canine teeth—Increased size and altered shape of the skull—Nakedness—Absence of a tail—Defenceless condition of man.

We have seen in the first chapter that the homological structure of man, his embryological development and the rudiments which he still retains, all declare in the plainest manner that he is descended from some lower form. The possession of exalted mental powers is no insuperable objection to this conclusion. In order that an ape-like creature should have been transformed into man, it is necessary that this early form, as well as many successive links, should all have varied in mind and body. It is impossible to obtain direct evidence on this head; but if it can be shewn that man now varies—that his variations are induced by the same general causes, and obey the same general laws, as in the case of the lower animals—there can be little doubt that the preceding intermediate links varied in a like manner. The variations at each successive stage of descent must, also, have been in some manner accumulated and fixed.

108The facts and conclusions to be given in this chapter relate almost exclusively to the probable means by which the transformation of man has been effected, as far as his bodily structure is concerned. The following chapter will be devoted to the development of his intellectual and moral faculties. But the present discussion likewise bears on the origin of the different races or species of mankind, whichever term may be preferred.

It is manifest that man is now subject to much variability. No two individuals of the same race are quite alike. We may compare millions of faces, and each will be distinct. There is an equally great amount of diversity in the proportions and dimensions of the various parts of the body; the length of the legs being one of the most variable points.140 Although in some quarters of the world an elongated skull, and in other quarters a short skull prevails, yet there is great diversity of shape even within the limits of the same race, as with the aborigines of America and South Australia,—the latter a race "probably as pure and homogeneous in blood, customs, and language as any in existence"—and even with the inhabitants of so confined an area as the Sandwich Islands.141 An eminent dentist assures me that there is nearly as much diversity in the teeth, as in the features. The chief arteries so frequently run in abnormal courses, that it has been found useful for surgical purposes to calculate 109 from 12,000 corpses how often each course prevails.142 The muscles are eminently variable: thus those of the foot were found by Prof. Turner143 not to be strictly alike in any two out of fifty bodies; and in some the deviations were considerable. Prof. Turner adds that the power of performing the appropriate movements must have been modified in accordance with the several deviations. Mr. J. Wood has recorded144 the occurrence of 295 muscular variations in thirty-six subjects, and in another set of the same number no less than 558 variations, reckoning both sides of the body as one. In the last set, not one body out of the thirty-six was "found totally wanting in departures from the standard descriptions of the muscular system given in anatomical text-books." A single body presented the extraordinary number of twenty-five distinct abnormalities. The same muscle sometimes varies in many ways: thus Prof. Macalister describes145 no less than twenty distinct variations in the palmaris accessorius.

The famous old anatomist, Wolff,146 insists that the internal viscera are more variable than the external parts: Nulla particula est quæ non aliter et aliter in aliis se habeat hominibus. He has even written a treatise on the choice of typical examples of the viscera for representation. A discussion on the beau-ideal of the liver, lungs, kidneys, &c., as of the human face divine, sounds strange in our ears.

The variability or diversity of the mental faculties in men of the same race, not to mention the greater 110 differences between the men of distinct races, is so notorious that not a word need here be said. So it is with the lower animals, as has been illustrated by a few examples in the last chapter. All who have had charge of menageries admit this fact, and we see it plainly in our dogs and other domestic animals. Brehm especially insists that each individual monkey of those which he kept under confinement in Africa had its own peculiar disposition and temper: he mentions one baboon remarkable for its high intelligence; and the keepers in the Zoological Gardens pointed out to me a monkey, belonging to the New World division, equally remarkable for intelligence. Rengger, also, insists on the diversity in the various mental characters of the monkeys of the same species which he kept in Paraguay; and this diversity, as he adds, is partly innate, and partly the result of the manner in which they have been treated or educated.147

I have elsewhere148 so fully discussed the subject of Inheritance that I need here add hardly anything. A greater number of facts have been collected with respect to the transmission of the most trifling, as well as of the most important characters in man than in any of the lower animals; though the facts are copious enough with respect to the latter. So in regard to mental qualities, their transmission is manifest in our dogs, horses, and other domestic animals. Besides special tastes and habits, general intelligence, courage, bad and good temper, &c., are certainly transmitted. With man we see similar facts in almost every family; and we 111now know through the admirable labours of Mr. Galton149 that genius, which implies a wonderfully complex combination of high faculties, tends to be inherited; and, on the other hand, it is too certain that insanity and deteriorated mental powers likewise run in the same families.

With respect to the causes of variability we are in all cases very ignorant; but we can see that in man as in the lower animals, they stand in some relation with the conditions to which each species has been exposed during several generations. Domesticated animals vary more than those in a state of nature; and this is apparently due to the diversified and changing nature of their conditions. The different races of man resemble in this respect domesticated animals, and so do the individuals of the same race when inhabiting a very wide area, like that of America. We see the influence of diversified conditions in the more civilised nations, the members of which belong to different grades of rank and follow different occupations, presenting a greater range of character than the members of barbarous nations. But the uniformity of savages has often been exaggerated, and in some cases can hardly be said to exist.150 It is nevertheless an error to speak of man, even if we look only to the conditions to which he has been subjected, as "far more domesticated"151 than 112any other animal. Some savage races, such as the Australians, are not exposed to more diversified conditions than are many species which have very wide ranges. In another and much more important respect, man differs widely from any strictly domesticated animal; for his breeding has not been controlled, either through methodical or unconscious selection. No race or body of men has been so completely subjugated by other men, that certain individuals have been preserved and thus unconsciously selected, from being in some way more useful to their masters. Nor have certain male and female individuals been intentionally picked out and matched, except in the well-known case of the Prussian grenadiers; and in this case man obeyed, as might have been expected, the law of methodical selection; for it is asserted that many tall men were reared in the villages inhabited by the grenadiers with their tall wives.

If we consider all the races of man, as forming a single species, his range is enormous; but some separate races, as the Americans and Polynesians, have very wide ranges. It is a well-known law that widely-ranging species are much more variable than species with restricted ranges; and the variability of man may with more truth be compared with that of widely-ranging species, than with that of domesticated animals.

Not only does variability appear to be induced in man and the lower animals by the same general causes, but in both the same characters are affected in a closely analogous manner. This has been proved in such full detail by Godron and Quatrefages, that I need here only refer to their works.152 Monstrosities, which gra113duate into slight variations, are likewise so similar in man and the lower animals, that the same classification and the same terms can be used for both, as may be seen in Isidore Geoffroy St.-Hilaire's great work.153 This is a necessary consequence of the same laws of change prevailing throughout the animal kingdom. In my work on the variation of domestic animals, I have attempted to arrange in a rude fashion the laws of variation under the following heads:—The direct and definite action of changed conditions, as shewn by all or nearly all the individuals of the same species varying in the same manner under the same circumstances. The effects of the long-continued use or disuse of parts. The cohesion of homologous parts. The variability of multiple parts. Compensation of growth; but of this law I have found no good instances in the case of man. The effects of the mechanical pressure of one part on another; as of the pelvis on the cranium of the infant in the womb. Arrests of development, leading to the diminution or suppression of parts. The reappearance of long-lost characters through reversion. And lastly, correlated variation. All these so-called laws apply equally to man and the lower animals; and most of them even to plants. It would be superfluous here to discuss all of them;154 but several are so important for us, that they must be treated at considerable length.

The direct and definite action of changed conditions.—This is a most perplexing subject. It cannot be denied 114that changed conditions produce some effect, and occasionally a considerable effect, on organisms of all kinds; and it seems at first probable that if sufficient time were allowed this would be the invariable result. But I have failed to obtain clear evidence in favour of this conclusion; and valid reasons may be urged on the other side, at least as far as the innumerable structures are concerned, which are adapted for special ends. There can, however, be no doubt that changed conditions induce an almost indefinite amount of fluctuating variability, by which the whole organisation is rendered in some degree plastic.

In the United States, above 1,000,000 soldiers, who served in the late war, were measured, and the States in which they were born and reared recorded.155 From this astonishing number of observations it is proved that local influences of some kind act directly on stature; and we further learn that "the State where the physical growth has in great measure taken place, and the State of birth, which indicates the ancestry, seem to exert a marked influence on the stature." For instance it is established, "that residence in the Western States, during the years of growth, tends to produce increase of stature." On the other hand, it is certain that with sailors, their manner of life delays growth, as shewn "by the great difference between the statures of soldiers and sailors at the ages of 17 and 18 years." Mr. B. A. Gould endeavoured to ascertain the nature of the influences which thus act on stature; but he arrived only at negative results, namely, that they did not relate to climate, the elevation of the land, soil, nor even "in any controlling degree" to the abundance or need of the com115forts of life. This latter conclusion is directly opposed to that arrived at by Villermé from the statistics of the height of the conscripts in different parts of France. When we compare the differences in stature between the Polynesian chiefs and the lower orders within the same islands, or between the inhabitants of the fertile volcanic and low barren coral islands of the same ocean,156 or again between the Fuegians on the eastern and western shores of their country, where the means of subsistence are very different, it is scarcely possible to avoid the conclusion that better food and greater comfort do influence stature. But the preceding statements shew how difficult it is to arrive at any precise result. Dr. Beddoe has lately proved that, with the inhabitants of Britain, residence in towns and certain occupations have a deteriorating influence on height; and he infers that the result is to a certain extent inherited, as is likewise the case in the United States. Dr. Beddoe further believes that wherever a "race attains its maximum of physical development, it rises highest in energy and moral vigour."157

Whether external conditions produce any other direct effect on man is not known. It might have been expected that differences of climate would have had a marked influence, as the lungs and kidneys are brought into fuller activity under a low temperature, and the liver and skin under a high one.158 It was formerly thought that the colour of the skin and the character 116of the hair were determined by light or heat; and although it can hardly be denied that some effect is thus produced, almost all observers now agree that the effect has been very small, even after exposure during many ages. But this subject will be more properly discussed when we treat of the different races of mankind. With our domestic animals there are grounds for believing that cold and damp directly affect the growth of the hair; but I have not met with any evidence on this head in the case of man.

Effects of the increased Use and Disuse of Parts.—It is well known that use strengthens the muscles in the individual, and complete disuse, or the destruction of the proper nerve, weakens them. When the eye is destroyed the optic nerve often becomes atrophied. When an artery is tied, the lateral channels increase not only in diameter, but in the thickness and strength of their coats. When one kidney ceases acting from disease, the other increases in size and does double work. Bones increase not only in thickness, but in length, from carrying a greater weight.[159] Different occupations habitually followed lead to changed proportions in various parts of the body. Thus it was clearly ascertained by the United States Commission[160] that the legs of the sailors employed in the late war were longer by 0.217 of an inch than those of the soldiers, though the sailors were on an average shorter men; whilst their arms were shorter by 1.09 of an inch, and therefore out of proportion shorter in relation to 117their lesser height. This shortness of the arms is apparently due to their greater use, and is an unexpected result; but sailors chiefly use their arms in pulling and not in supporting weights. The girth of the neck and the depth of the instep are greater, whilst the circumference of the chest, waist, and hips is less in sailors than in soldiers.

Whether the several foregoing modifications would become hereditary, if the same habits of life were followed during many generations, is not known, but is probable. Rengger[161]attributes the thin legs and thick arms of the Payaguas Indians to successive generations having passed nearly their whole lives in canoes, with their lower extremities motionless. Other writers have come to a similar conclusion in other analogous cases. According to Cranz,[162] who lived for a long time with the Esquimaux, "the natives believe that ingenuity and dexterity in seal-catching (their highest art and virtue) is hereditary; there is really something in it, for the son of a celebrated seal-catcher will distinguish himself though he lost his father in childhood." But in this case it is mental aptitude, quite as much as bodily structure, which appears to be inherited. It is asserted that the hands of English labourers are at birth larger than those of the gentry.[163] From the correlation which exists, at least in some cases,[164] between the development of the extremities and of the jaws, it is possible that in those classes which do not labour much with their hands and feet, the jaws would be reduced in size from this cause. That they are generally smaller in refined and civilised men than in hard-working men or savages, 118is certain. But with savages, as Mr. Herbert Spencer[165] has remarked, the greater use of the jaws in chewing coarse, uncooked food, would act in a direct manner on the masticatory muscles and on the bones to which they are attached. In infants long before birth, the skin on the soles of the feet is thicker than on any other part of the body;[166] and it can hardly be doubted that this is due to the inherited effects of pressure during a long series of generations.

It is familiar to every one that watchmakers and engravers are liable to become short-sighted, whilst sailors and especially savages are generally long-sighted. Short-sight and long-sight certainly tend to be inherited.[167] The inferiority of Europeans, in comparison with savages, in eyesight and in the other senses, is no doubt the accumulated and transmitted effect of lessened use during many generations; for Rengger[168] states that he has repeatedly observed Europeans, who had been brought up and spent their whole lives with the wild Indians, who nevertheless did not equal them in the sharpness of their senses. The same naturalist observes that the cavities in the skull for the reception of the several sense-organs are larger in the American aborigines than in Europeans; and this no doubt indicates a corresponding difference in the dimensions of the organs themselves. Blumenbach has also remarked on the large size of the nasal cavities 119in the skulls of the American aborigines, and connects this fact with their remarkably acute power of smell. The Mongolians of the plains of Northern Asia, according to Pallas, have wonderfully perfect senses; and Prichard believes that the great breadth of their skulls across the zygomas follows from their highly-developed sense-organs.169

The Quechua Indians inhabit the lofty plateaux of Peru, and Alcide d'Orbigny states170 that from continually breathing a highly rarefied atmosphere they have acquired chests and lungs of extraordinary dimensions. The cells, also, of the lungs are larger and more numerous than in Europeans. These observations have been doubted; but Mr. D. Forbes carefully measured many Aymaras, an allied race, living at the height of between ten and fifteen thousand feet; and he informs me171 that they differ conspicuously from the men of all other races seen by him, in the circumference and length of their bodies. In his table of measurements, the stature of each man is taken at 1000, and the other measurements are reduced to this standard. It is here seen that the extended arms of the Aymaras are shorter than those of Europeans, and much shorter than those of Negroes. The legs are likewise shorter, and they present this remarkable peculiarity, that in every Aymara measured the femur is actually shorter than the tibia. On an average the length of the femur to that of the tibia is as 211 to 252; whilst in two Europeans measured at the same 120time, the femora to the tibiæ were as 244 to 230; and in three Negroes as 258 to 241. The humerus is likewise shorter relatively to the fore-arm. This shortening of that part of the limb which is nearest to the body, appears to be, as suggested to me by Mr. Forbes, a case of compensation in relation with the greatly increased length of the trunk. The Aymaras present some other singular points of structure, for instance, the very small projection of the heel.

These men are so thoroughly acclimatised to their cold and lofty abode, that when formerly carried down by the Spaniards to the low Eastern plains, and when now tempted down by high wages to the gold-washings, they suffer a frightful rate of mortality. Nevertheless Mr. Forbes found a few pure families which had survived during two generations; and he observed that they still inherited their characteristic peculiarities. But it was manifest, even without measurement, that these peculiarities had all decreased; and on measurement their bodies were found not to be so much elongated as those of the men on the high plateau; whilst their femora had become somewhat lengthened, as had their tibiæ but in a less degree. The actual measurements may be seen by consulting Mr. Forbes' memoir. From these valuable observations, there can, I think, be no doubt that residence during many generations at a great elevation tends, both directly and indirectly, to induce inherited modifications in the proportions of the body.172

Although man may not have been much modified during the latter stages of his existence through the 121increased or decreased use of parts, the facts now given shew that his liability in this respect has not been lost; and we positively know that the same law holds good with the lower animals. Consequently we may infer, that when at a remote epoch the progenitors of man were in a transitional state, and were changing from quadrupeds into bipeds, natural selection would probably have been greatly aided by the inherited effects of the increased or diminished use of the different parts of the body.

Arrests of Development.—Arrested development differs from arrested growth, as parts in the former state continue to grow whilst still retaining their early condition. Various monstrosities come under this head, and some are known to be occasionally inherited, as a cleft-palate. It will suffice for our purpose to refer to the arrested brain-development of microcephalous idiots, as described in Vogt's great memoir.173 Their skulls are smaller, and the convolutions of the brain are less complex than in normal men. The frontal sinus, or the projection over the eyebrows, is largely developed, and the jaws are prognathous to an "effrayant" degree; so that these idiots somewhat resemble the lower types of mankind. Their intelligence and most of their mental faculties are extremely feeble. They cannot acquire the power of speech, and are wholly incapable of prolonged attention, but are much given to imitation. They are strong and remarkably active, continually gamboling and jumping about, and making grimaces. They often ascend stairs on all-fours; and are curiously fond of climbing up furniture or trees. We are thus reminded of the delight 122shewn by almost all boys in climbing trees; and this again reminds us how lambs and kids, originally alpine animals, delight to frisk on any hillock, however small.

Reversion.—Many of the cases to be here given might have been introduced under the last heading. Whenever a structure is arrested in its development, but still continues growing until it closely resembles a corresponding structure in some lower and adult member of the same group, we may in one sense consider it as a case of reversion. The lower members in a group give us some idea how the common progenitor of the group was probably constructed; and it is hardly credible that a part arrested at an early phase of embryonic development should be enabled to continue growing so as ultimately to perform its proper function, unless it had acquired this power of continued growth during some earlier state of existence, when the present exceptional or arrested structure was normal. The simple brain of a microcephalous idiot, in as far as it resembles that of an ape, may in this sense be said to offer a case of reversion. There are other cases which come more strictly under our present heading of reversion. Certain structures, regularly occurring in the lower members of the group to which man belongs, occasionally make their appearance in him, though not found in the normal human embryo; or, if present in the normal human embryo, they become developed in an abnormal manner, though this manner of development is proper to the lower members of the same group. These remarks will be rendered clearer by the following illustrations.

In various mammals the uterus graduates from a double organ with two distinct orifices and two passages, as in the marsupials, into a single organ, showing no signs of doubleness except a slight internal fold, as in123 the higher apes and man. The rodents exhibit a perfect series of gradations between these two extreme states. In all mammals the uterus is developed from two simple primitive tubes, the inferior portions of which form the cornua; and it is in the words of Dr. Farre "by the coalescence of the two cornua at their lower extremities that the body of the uterus is formed in man; while in those animals in which no middle portion or body exists, the cornua remain ununited. As the development of the uterus proceeds, the two cornua become gradually shorter, until at length they are lost, or, as it were, absorbed into the body of the uterus." The angles of the uterus are still produced into cornua, even so high in the scale as in the lower apes, and their allies the lemurs.

Now in women anomalous cases are not very infrequent, in which the mature uterus is furnished with cornua, or is partially divided into two organs; and such cases, according to Owen, repeat "the grade of concentrative development," attained by certain rodents. Here perhaps we have an instance of a simple arrest of embryonic development, with subsequent growth and perfect functional development, for either side of the partially double uterus is capable of performing the proper office of gestation. In other and rarer cases, two distinct uterine cavities are formed, each having its proper orifice and passage.174 No such stage is passed through during the ordinary development of the embryo, and it is difficult to believe, though perhaps not impossible, that the two simple, minute, primitive tubes could know how (if such an expression may be used) to 124grow into two distinct uteri, each with a well-constructed orifice and passage, and each furnished with numerous muscles, nerves, glands and vessels, if they had not formerly passed through a similar course of development, as in the case of existing marsupials. No one will pretend that so perfect a structure as the abnormal double uterus in woman could be the result of mere chance. But the principle of reversion, by which long-lost dormant structures are called back into existence, might serve as the guide for the full development of the organ, even after the lapse of an enormous interval of time.

Professor Canestrini,175 after discussing the foregoing and various analogous cases, arrives at the same conclusion as that just given. He adduces, as another instance, the malar bone, which, in some of the Quadrumana and other mammals, normally consists of two portions. This is its condition in the two-months-old human fœtus; and thus it sometimes remains, through arrested development, in man when adult, more especially in the lower prognathous races. Hence Canestrini concludes that some ancient progenitor of man must have possessed this bone normally divided into two portions, which subsequently became fused together. In man the frontal bone consists of a single piece, but in the embryo and in children, and in almost all the lower mammals, it consists of two pieces separated by a distinct suture. This suture occasionally persists, more or less distinctly, in man after maturity, and more fre125quently in ancient than in recent crania, especially as Canestrini has observed in those exhumed from the Drift and belonging to the brachycephalic type. Here again he comes to the same conclusion as in the analogous case of the malar bones. In this and other instances presently to be given, the cause of ancient races approaching the lower animals in certain characters more frequently than do the modern races, appears to be that the latter stand at a somewhat greater distance in the long line of descent from their early semi-human progenitors.

Various other anomalies in man, more or less analogous with the foregoing, have been advanced by different authors176 as cases of reversion; but these seem not a little doubtful, for we have to descend extremely low in the mammalian series before we find such structures normally present.177

126

In man the canine teeth are perfectly efficient instruments for mastication. But their true canine character, as Owen178 remarks, "is indicated by the conical form of the crown, which terminates in an obtuse point, is convex outward and flat or sub-concave within, at the base of which surface there is a feeble prominence. The conical form is best expressed in the Melanian races, especially the Australian. The canine is more deeply implanted, and by a stronger fang than the incisors." Nevertheless this tooth no longer serves man as a special weapon for tearing his enemies or prey; it may, therefore, as far as its proper function is concerned, be considered as rudimentary. In every large collection of human skulls some may be found, as Häckel179 observes, with the canine teeth projecting considerably beyond the others in the same manner, but in a less degree, as in the anthropomorphous apes. In these cases, open spaces between the teeth in the one jaw are left for the reception of the canines belonging to the opposite jaw. An interspace of this kind in a Kaffir skull, figured by Wagner, is surprisingly wide.180 Considering how few ancient skulls have been examined in comparison with recent skulls, it is an interesting fact that in at least three cases the canines project largely; and in the Naulette jaw they are spoken of as enormous.181

The males alone of the anthropomorphous apes have their canines fully developed; but in the female gorilla, and in a less degree in the female orang, these teeth project considerably beyond the others; therefore the fact that women sometimes have, as I have been assured, considerably projecting canines, is no serious objection to the belief that their occasional great development in man is a case of reversion to an ape-like progenitor. He who rejects with scorn the belief that the shape of his own canines, and their occasional great development in other men, are due to our early progenitors having been provided with these formidable weapons, will probably reveal by sneering the line of his descent. For though he no longer intends, nor has the power, to use these teeth as weapons, he will unconsciously retract his "snarling muscles" (thus named by Sir C. Bell)[182] so as to expose them ready for action, like a dog prepared to fight.

Many muscles are occasionally developed in man, which are proper to the Quadrumana or other mammals. Professor Vlacovich[183] examined forty male subjects, and found a muscle, called by him the ischiopubic, in nineteen of them; in three others there was a ligament which represented this muscle; and in the remaining eighteen no trace of it. Out of thirty female subjects this muscle was developed on both sides in only two, but in three others the rudimentary ligament was present. This muscle, therefore, appears to be much more common in the male than in the female sex; and on the principle of the descent of man from some lower form, its presence can be understood; for it has been detected in several of the lower animals, and in all of 128these it serves exclusively to aid the male in the act of reproduction.

Mr. J. Wood, in his valuable series of papers,[184] has minutely described a vast number of muscular variations in man, which resemble normal structures in the lower animals. Looking only to the muscles which closely resemble those regularly present in our nearest allies, the Quadrumana, they are too numerous to be here even specified. In a single male subject, having a strong bodily frame and well-formed skull, no less than seven muscular variations were observed, all of which plainly represented muscles proper to various kinds of apes. This man, for instance, had on both sides of his neck a true and powerful "levator claviculæ," such as is found in all kinds of apes, and which is said to occur in about one out of sixty human subjects.[185] Again, this man had "a special abductor of the metatarsal bone of the fifth digit, such as Professor Huxley and Mr. Flower have shewn to exist uniformly in the higher and lower apes." The hands and arms of man are eminently characteristic structures, but their muscles are extremely liable to vary, so as to resemble the corresponding muscles in the lower animals.[186] Such resemblances are either complete and per129fect or imperfect, yet in this latter case manifestly of a transitional nature. Certain variations are more common in man, and others in woman, without our being able to assign any reason. Mr. Wood, after describing numerous cases, makes the following pregnant remark: "Notable departures from the ordinary type of the muscular structures run in grooves or directions, which must be taken to indicate some unknown factor, of much importance to a comprehensive knowledge of general and scientific anatomy."[187]

That this unknown factor is reversion to a former state of existence may be admitted as in the highest degree probable. It is quite incredible that a man should through mere accident abnormally resemble, in no less than seven of his muscles, certain apes, if there had been no genetic connection between them. On the other hand, if man is descended from some ape-like creature, no valid reason can be assigned why certain muscles should not suddenly reappear after an interval of many thousand generations, in the same manner as with horses, asses, and mules, dark-coloured stripes suddenly reappear on the legs and shoulders, after an interval of hundreds, or more probably thousands, of generations.

These various cases of reversion are so closely related 130to those of rudimentary organs given in the first chapter, that many of them might have been indifferently introduced in either chapter. Thus a human uterus furnished with cornua may be said to represent in a rudimentary condition the same organ in its normal state in certain mammals. Some parts which are rudimental in man, as the os coccyx in both sexes and the mammæ in the male sex, are always present; whilst others, such as the supra-condyloid foramen, only occasionally appear, and therefore might have been introduced under the head of reversion. These several reversionary, as well as the strictly rudimentary, structures reveal the descent of man from some lower form in an unmistakeable manner.

Correlated Variation.—In man, as in the lower animals, many structures are so intimately related, that when one part varies so does another, without our being able, in most cases, to assign any reason. We cannot say whether the one part governs the other, or whether both are governed by some earlier developed part. Various monstrosities, as I. Geoffroy repeatedly insists, are thus intimately connected. Homologous structures are particularly liable to change together, as we see on the opposite sides of the body, and in the upper and lower extremities. Meckel long ago remarked that when the muscles of the arm depart from their proper type, they almost always imitate those of the leg; and so conversely with the muscles of the legs. The organs of sight and hearing, the teeth and hair, the colour of the skin and hair, colour and constitution, are more or less correlated.188 Professor Schaaffhausen first drew attention to the rela131tion apparently existing between a muscular frame and strongly-pronounced supra-orbital ridges, which are so characteristic of the lower races of man.

Besides the variations which can be grouped with more or less probability under the foregoing heads, there is a large class of variations which may be provisionally called spontaneous, for they appear, owing to our ignorance, to arise without any exciting cause. It can, however, be shewn that such variations, whether consisting of slight individual differences, or of strongly-marked and abrupt deviations of structure, depend much more on the constitution of the organism than on the nature of the conditions to which it has been subjected.189

Rate of Increase.—Civilised populations have been known under favourable conditions, as in the United States, to double their number in twenty-five years; and according to a calculation by Euler, this might occur in a little over twelve years.190 At the former rate the present population of the United States, namely, thirty millions, would in 657 years cover the whole terraqueous globe so thickly, that four men would have to stand on each square yard of surface. The primary or fundamental check to the continued increase of man is the difficulty of gaining subsistence and of living in comfort. We may infer that this is the case from what we see, for instance, in the United States, where subsistence is easy and there is plenty of room. If such means were suddenly doubled in Great Britain, our number would be quickly doubled. With civilised nations the 132above primary check acts chiefly by restraining marriages. The greater death-rate of infants in the poorest classes is also very important; as well as the greater mortality at all ages, and from various diseases, of the inhabitants of crowded and miserable houses. The effects of severe epidemics and wars are soon counterbalanced, and more than counterbalanced, in nations placed under favourable conditions. Emigration also comes in aid as a temporary check, but not to any great extent with the extremely poor classes.

There is reason to suspect, as Malthus has remarked, that the reproductive power is actually less in barbarous than in civilised races. We know nothing positively on this head, for with savages no census has been taken; but from the concurrent testimony of missionaries, and of others who have long resided with such people, it appears that their families are usually small, and large ones rare. This may be partly accounted for, as it is believed, by the women suckling their infants for a prolonged period; but it is highly probable that savages, who often suffer much hardship, and who do not obtain so much nutritious food as civilised men, would be actually less prolific. I have shewn in a former work,[191] that all our domesticated quadrupeds and birds, and all our cultivated plants, are more fertile than the corresponding species in a state of nature. It is no valid objection to this conclusion that animals suddenly supplied with an excess of food, or when rendered very fat, and that most plants when suddenly removed from very poor to very rich soil, are rendered more or less sterile. We might, therefore, expect that civilised men, who in one sense are highly domesticated, would 133be more prolific than wild men. It is also probable that the increased fertility of civilised nations would become, as with our domestic animals, an inherited character: it is at least known that with mankind a tendency to produce twins runs in families.[192]

Notwithstanding that savages appear to be less prolific than civilised people, they would no doubt rapidly increase if their numbers were not by some means rigidly kept down. The Santali, or hill-tribes of India, have recently afforded a good illustration of this fact; for they have increased, as shewn by Mr. Hunter,[193] at an extraordinary rate since vaccination has been introduced, other pestilences mitigated, and war sternly repressed. This increase, however, would not have been possible had not these rude people spread into the adjoining districts and worked for hire. Savages almost always marry; yet there is some prudential restraint, for they do not commonly marry at the earliest possible age. The young men are often required to show that they can support a wife, and they generally have first to earn the price with which to purchase her from her parents. With savages the difficulty of obtaining subsistence occasionally limits their number in a much more direct manner than with civilised people, for all tribes periodically suffer from severe famines. At such times savages are forced to devour much bad food, and their health can hardly fail to be injured. Many accounts have been published of their protruding stomachs and emaciated limbs after and during famines. They are then, also, compelled to wander much about, and their infants, as I was assured in Australia, perish 134in large numbers. As famines are periodical, depending chiefly on extreme seasons, all tribes must fluctuate in number. They cannot steadily and regularly increase, as there is no artificial increase in the supply of food. Savages when hardly pressed encroach on each other's territories, and war is the result; but they are indeed almost always at war with their neighbours. They are liable to many accidents on land and water in their search for food; and in some countries they must suffer much from the larger beasts of prey. Even in India, districts have been depopulated by the ravages of tigers.

Malthus has discussed these several checks, but he does not lay stress enough on what is probably the most important of all, namely infanticide, especially of female infants, and the habit of procuring abortion. These practices now prevail in many quarters of the world, and infanticide seems formerly to have prevailed, as Mr. M'Lennan[194] has shewn, on a still more extensive scale. These practices appear to have originated in savages recognising the difficulty, or rather the impossibility of supporting all the infants that are born. Licentiousness may also be added to the foregoing checks; but this does not follow from failing means of subsistence; though there is reason to believe that in some cases (as in Japan) it has been intentionally encouraged as a means of keeping down the population.

If we look back to an extremely remote epoch, before man had arrived at the dignity of manhood, he would have been guided more by instinct and less by reason than are savages at the present time. Our early semi-human progenitors would not have practised infanticide, for the instincts of the lower animals are never so perverted as to lead them regularly to destroy their own 135offspring. There would have been no prudential restraint from marriage, and the sexes would have freely united at an early age. Hence the progenitors of man would have tended to increase rapidly, but checks of some kind, either periodical or constant, must have kept down their numbers, even more severely than with existing savages. What the precise nature of these checks may have been, we cannot say, any more than with most other animals. We know that horses and cattle, which are not highly prolific animals, when first turned loose in South America, increased at an enormous rate. The slowest breeder of all known animals, namely the elephant, would in a few thousand years stock the whole world. The increase of every species of monkey must be checked by some means; but not, as Brehm remarks, by the attacks of beasts of prey. No one will assume that the actual power of reproduction in the wild horses and cattle of America, was at first in any sensible degree increased; or that, as each district became fully stocked, this same power was diminished. No doubt in this case and in all others, many checks concur, and different checks under different circumstances; periodical dearths, depending on unfavourable seasons, being probably the most important of all. So it will have been with the early progenitors of man.

Natural Selection.—We have now seen that man is variable in body and mind; and that the variations are induced, either directly or indirectly, by the same general causes, and obey the same general laws, as with the lower animals. Man has spread widely over the face of the earth, and must have been exposed, during his incessant migrations,[195] to the most diversified con136ditions. The inhabitants of Tierra del Fuego, the Cape of Good Hope, and Tasmania in the one hemisphere, and of the Arctic regions in the other, must have passed through many climates and changed their habits many times, before they reached their present homes.[196] The early progenitors of man must also have tended, like all other animals, to have increased beyond their means of subsistence; they must therefore occasionally have been exposed to a struggle for existence, and consequently to the rigid law of natural selection. Beneficial variations of all kinds will thus, either occasionally or habitually, have been preserved, and injurious ones eliminated. I do not refer to strongly-marked deviations of structure, which occur only at long intervals of time, but to mere individual differences. We know, for instance, that the muscles of our hands and feet, which determine our powers of movement, are liable, like those of the lower animals,[197] to incessant variability. If then the ape-like progenitors of man which inhabited any district, especially one undergoing some change in its conditions, were divided into two equal bodies, the one half which included all the individuals best adapted by their powers of movement for gaining subsistence or for defending themselves, would on an average survive in greater number and procreate more offspring than the other and less well endowed half.

Man in the rudest state in which he now exists is the most dominant animal that has ever appeared on the earth. He has spread more widely than any 137other highly organised form; and all others have yielded before him. He manifestly owes this immense superiority to his intellectual faculties, his social habits, which lead him to aid and defend his fellows, and to his corporeal structure. The supreme importance of these characters has been proved by the final arbitrament of the battle for life. Through his powers of intellect, articulate language has been evolved; and on this his wonderful advancement has mainly depended. He has invented and is able to use various weapons, tools, traps, &c., with which he defends himself, kills or catches prey, and otherwise obtains food. He has made rafts or canoes on which to fish or cross over to neighbouring fertile islands. He has discovered the art of making fire, by which hard and stringy roots can be rendered digestible, and poisonous roots or herbs innocuous. This last discovery, probably the greatest, excepting language, ever made by man, dates from before the dawn of history. These several inventions, by which man in the rudest state has become so preeminent, are the direct result of the development of his powers of observation, memory, curiosity, imagination, and reason. I cannot, therefore, understand how it is that Mr. Wallace[198] maintains, that "natural selection could only have endowed the savage with a brain a little superior to that of an ape."

138

Although the intellectual powers and social habits of man are of paramount importance to him, we must not underrate the importance of his bodily structure, to which subject the remainder of this chapter will be devoted. The development of the intellectual and social or moral faculties will be discussed in the following chapter.

Even to hammer with precision is no easy matter, as every one who has tried to learn carpentry will admit. To throw a stone with as true an aim as can a Fuegian in defending himself, or in killing birds, requires the most consummate perfection in the correlated action of the muscles of the hand, arm, and shoulder, not to mention a fine sense of touch. In throwing a stone or spear, and in many other actions, a man must stand firmly on his feet; and this again demands the perfect coadaptation of numerous muscles. To chip a flint into the rudest tool, or to form a barbed spear or hook from a bone, demands the use of a perfect hand; for, as a most capable judge, Mr. Schoolcraft,[199] remarks, the shaping fragments of stone into knives, lances, or arrow-heads, shews "extraordinary ability and long practice." We have evidence of this in primeval men having practised a division of labour; each man did not manufacture his own flint tools or rude pottery; but certain individuals appear to have devoted themselves to such work, no doubt receiving in exchange the produce of the chase. Archæologists are convinced that an enormous interval of time 139elapsed before our ancestors thought of grinding chipped flints into smooth tools. A man-like animal who possessed a hand and arm sufficiently perfect to throw a stone with precision or to form a flint into a rude tool, could, it can hardly be doubted, with sufficient practice make almost anything, as far as mechanical skill alone is concerned, which a civilised man can make. The structure of the hand in this respect may be compared with that of the vocal organs, which in the apes are used for uttering various signal-cries, or, as in one species, musical cadences; but in man closely similar vocal organs have become adapted through the inherited effects of use for the utterance of articulate language.

Turning now to the nearest allies of man, and therefore to the best representatives of our early progenitors, we find that the hands in the Quadrumana are constructed on the same general pattern as in us, but are far less perfectly adapted for diversified uses. Their hands do not serve so well as the feet of a dog for locomotion; as may be seen in those monkeys which walk on the outer margins of the palms, or on the backs of their bent fingers, as in the chimpanzee and orang.[200] Their hands, however, are admirably adapted for climbing trees. Monkeys seize thin branches or ropes, with the thumb on one side and the fingers and palm on the other side, in the same manner as we do. They can thus also carry rather large objects, such as the neck of a bottle, to their mouths. Baboons turn over stones and scratch up roots with their hands. They seize nuts, insects, or other small objects with the thumb in opposition to the fingers, and no doubt they thus extract eggs and the young from the nests of birds. American monkeys beat the wild oranges on the 140branches until the rind is cracked, and then tear it off with the fingers of the two hands. Other monkeys open mussel-shells with the two thumbs. With their fingers they pull out thorns and burrs, and hunt for each other's parasites. In a state of nature they break open hard fruits with the aid of stones. They roll down stones or throw them at their enemies; nevertheless, they perform these various actions clumsily, and they are quite unable, as I have myself seen, to throw a stone with precision.

It seems to me far from true that because "objects are grasped clumsily" by monkeys, "a much less specialised organ of prehension" would have served them201 as well as their present hands. On the contrary, I see no reason to doubt that a more perfectly constructed hand would have been an advantage to them, provided, and it is important to note this, that their hands had not thus been rendered less well adapted for climbing trees. We may suspect that a perfect hand would have been disadvantageous for climbing; as the most arboreal monkeys in the world, namely Ateles in America and Hylobates in Asia, either have their thumbs much reduced in size and even rudimentary, or their fingers partially coherent, so that their hands are converted into mere grasping-hooks.202

As soon as some ancient member in the great series of the Primates came, owing to a change in its manner of procuring subsistence, or to a change in the conditions of its native country, to live somewhat less on trees and more on the ground, its manner of progression would have been modified; and in this case it 141would have had to become either more strictly quadrupedal or bipedal. Baboons frequent hilly and rocky districts, and only from necessity climb up high trees;203 and they have acquired almost the gait of a dog. Man alone has become a biped; and we can, I think, partly see how he has come to assume his erect attitude, which forms one of the most conspicuous differences between him and his nearest allies. Man could not have attained his present dominant position in the world without the use of his hands which are so admirably adapted to act in obedience to his will. As Sir C. Bell204 insists "the hand supplies all instruments, and by its correspondence with the intellect gives him universal dominion." But the hands and arms could hardly have become perfect enough to have manufactured weapons, or to have hurled stones and spears with a true aim, as long as they were habitually used for locomotion and for supporting the whole weight of the body, or as long as they were especially well adapted, as previously remarked, for climbing trees. Such rough treatment would also have blunted the sense of touch, on which their delicate use largely depends. From these causes alone it would have been an advantage to man to have become a biped; but for many actions it is almost necessary that both arms and the whole upper part of the body should be free; and he must for this end stand firmly on his feet. To gain this great advantage, the feet have been rendered flat, and the great toe peculiarly modified, though this has entailed the loss of the power of prehension. It accords with the principle of the division of physiological labour, which prevails throughout the animal kingdom, that 142as the hands became perfected for prehension, the feet should have become perfected for support and locomotion. With some savages, however, the foot has not altogether lost its prehensile power, as shewn by their manner of climbing trees and of using them in other ways.205

If it be an advantage to man to have his hands and arms free and to stand firmly on his feet, of which there can be no doubt from his preeminent success in the battle of life, then I can see no reason why it should not have been advantageous to the progenitors of man to have become more and more erect or bipedal. They would thus have been better able to have defended themselves with stones or clubs, or to have attacked their prey, or otherwise obtained food. The best constructed individuals would in the long run have succeeded best, and have survived in larger numbers. If the gorilla and a few allied forms had become extinct, it might have been argued with great force and apparent truth, that an animal could not have been gradually converted from a quadruped into a biped; as all the individuals in an intermediate condition would have been miserably ill-fitted for progression. But we know (and this is well worthy of reflection) that several kinds of apes are now actually in this intermediate condition; and no one doubts that they are on the whole well adapted for their conditions of life. Thus the gorilla runs with a sidelong shambling gait, but more commonly 143progresses by resting on its bent hands. The long-armed apes occasionally use their arms like crutches, swinging their bodies forward between them, and some kinds of Hylobates, without having been taught, can walk or run upright with tolerable quickness; yet they move awkwardly, and much less securely than man. We see, in short, with existing monkeys various gradations between a form of progression strictly like that of a quadruped and that of a biped or man.

As the progenitors of man became more and more erect, with their hands and arms more and more modified for prehension and other purposes, with their feet and legs at the same time modified for firm support and progression, endless other changes of structure would have been necessary. The pelvis would have had to be made broader, the spine peculiarly curved and the head fixed in an altered position, and all these changes have been attained by man. Prof. Schaaffhausen206 maintains that "the powerful mastoid processes of the human skull are the result of his erect position;" and these processes are absent in the orang, chimpanzee, &c., and are smaller in the gorilla than in man. Various other structures might here have been specified, which appear connected with man's erect position. It is very difficult to decide how far all these correlated modifications are the result of natural selection, and how far of the inherited effects of the increased use of certain parts, or of the action of one part on another. No doubt these means of change act and react on each other: thus when certain muscles, and the crests of bone to which they are attached, become enlarged by 144habitual use, this shews that certain fictions are habitually performed and must be serviceable. Hence the individuals which performed them best, would tend to survive in greater numbers.

The free use of the arms and hands, partly the cause and partly the result of man's erect position, appears to have led in an indirect manner to other modifications of structure. The early male progenitors of man were, as previously stated, probably furnished with great canine teeth; but as they gradually acquired the habit of using stones, clubs, or other weapons, for fighting with their enemies, they would have used their jaws and teeth less and less. In this case, the jaws, together with the teeth, would have become reduced in size, as we may feel sure from innumerable analogous cases. In a future chapter we shall meet with a closely-parallel case, in the reduction or complete disappearance of the canine teeth in male ruminants, apparently in relation with the development of their horns; and in horses, in relation with their habit of fighting with their incisor teeth and hoofs.

In the adult male anthropomorphous apes, as Rütimeyer,207 and others have insisted, it is precisely the effect which the jaw-muscles by their great development have produced on the skull, that causes it to differ so greatly in many respects from that of man, and has given to it "a truly frightful physiognomy." Therefore as the jaws and teeth in the progenitors of man gradually become reduced in size, the adult skull would have presented nearly the same characters which it offers in the young of the anthropomorphous apes, and would thus have come to resemble more nearly that of existing 145man. A great reduction of the canine teeth in the males would almost certainly, as we shall hereafter see, have affected through inheritance the teeth of the females.

As the various mental faculties were gradually developed, the brain would almost certainly have become larger. No one, I presume, doubts that the large size of the brain in man, relatively to his body, in comparison with that of the gorilla or orang, is closely connected with his higher mental powers. We meet with closely analogous facts with insects, in which the cerebral ganglia are of extraordinary dimensions in ants; these ganglia in all the Hymenoptera being many times larger than in the less intelligent orders, such as beetles.208 On the other hand, no one supposes that the intellect of any two animals or of any two men can be accurately gauged by the cubic contents of their skulls. It is certain that there may be extraordinary mental activity with an extremely small absolute mass of nervous matter: thus the wonderfully diversified instincts, mental powers, and affections of ants are generally known, yet their cerebral ganglia are not so large as the quarter of a small pin's head. Under this latter point of view, the brain of an ant is one of the most marvellous atoms of matter in the world, perhaps more marvellous than the brain of man.

The belief that there exists in man some close relation between the size of the brain and the development of the intellectual faculties is supported by the comparison of the skulls of savage and civilised races, of ancient and modern people, and by the analogy of the whole verte146brate series. Dr. J. Barnard Davis has proved209 by many careful measurements, that the mean internal capacity of the skull in Europeans is 92·3 cubic inches; in Americans 87·5; in Asiatics 87·1; and in Australians only 81·9 inches. Professor Broca210 found that skulls from graves in Paris of the nineteenth century, were larger than those from vaults of the twelfth century, in the proportion of 1484 to 1426; and Prichard is persuaded that the present inhabitants of Britain have "much more capacious brain-cases" than the ancient inhabitants. Nevertheless it must be admitted that some skulls of very high antiquity, such as the famous one of Neanderthal, are well developed and capacious. With respect to the lower animals, M. E. Lartet,211 by comparing the crania of tertiary and recent mammals, belonging to the same groups, has come to the remarkable conclusion that the brain is generally larger and the convolutions more complex in the more recent form. On the other hand I have shewn212 that the brains of domestic rabbits are considerably reduced in bulk, in comparison with those of the wild rabbit or hare; and this may be attributed to their having been closely confined during many generations, so that they have exerted but little their intellect, instincts, senses, and voluntary movements.

The gradually increasing weight of the brain and skull in man must have influenced the development of the supporting spinal column, more especially whilst he was becoming erect. As this change of position was 147being brought about, the internal pressure of the brain, will, also, have influenced the form of the skull; for many facts shew how easily the skull is thus affected. Ethnologists believe that it is modified by the kind of cradle in which infants sleep. Habitual spasms of the muscles and a cicatrix from a severe burn have permanently modified the facial bones. In young persons whose heads from disease have become fixed either sideways or backwards, one of the eyes has changed its position, and the bones of the skull have been modified; and this apparently results from the brain pressing in a new direction.213 I have shewn that with long-eared rabbits, even so trifling a cause as the lopping forward of one ear drags forward on that side almost every bone of the skull; so that the bones on the opposite sides no longer strictly correspond. Lastly, if any animal were to increase or diminish much in general size, without any change in its mental powers; or if the mental powers were to be much increased or diminished without any great change in the size of the body; the shape of the skull would almost certainly be altered. I infer this from my observations on domestic rabbits, some kinds of which have become very much larger than the wild animal, whilst others have retained nearly the same size, but in both cases the brain has been much reduced relatively to the size of the body. Now I was at first much surprised by finding that in all these rabbits the skull had become elongated or dolicho148cephalic; for instance, of two skulls of nearly equal breadth, the one from a wild rabbit and the other from a large domestic kind, the former was only 3·15 and the latter 4·3 inches in length.214 One of the most marked distinctions in different races of man is that the skull in some is elongated, and in others rounded; and here the explanation suggested by the case of the rabbits may partially hold good; for Welcker finds that short "men incline more to brachycephaly, and tall men to dolichocephaly;"215 and tall men may be compared with the larger and longer-bodied rabbits, all of which have elongated skulls, or are dolichocephalic.

From these several facts we can to a certain extent understand the means through which the great size and more or less rounded form of the skull has been acquired by man; and these are characters eminently distinctive of him in comparison with the lower animals.

Another most conspicuous difference between man and the lower animals is the nakedness of his skin. Whales and dolphins (Cetacea), dugongs (Sirenia) and the hippopotamus are naked; and this may be advantageous to them for gliding through the water; nor would it be injurious to them from the loss of warmth, as the species which inhabit the colder regions are protected by a thick layer of blubber, serving the same purpose as the fur of seals and otters. Elephants and rhinoceroses are almost hairless; and as certain extinct species which formerly lived under an arctic climate were covered with long wool or hair, it would almost appear as if the existing species of both genera had lost 149their hairy covering from exposure to heat. This appears the more probable, as the elephants in India which live on elevated and cool districts are more hairy216 than those on the lowlands. May we then infer that man became divested of hair from having aboriginally inhabited some tropical land? The fact of the hair being chiefly retained in the male sex on the chest and face, and in both sexes at the junction of all four limbs with the trunk, favours this inference, assuming that the hair was lost before man became erect; for the parts which now retain most hair would then have been most protected from the heat of the sun. The crown of the head, however, offers a curious exception, for at all times it must have been one of the most exposed parts, yet it is thickly clothed with hair. In this respect man agrees with the great majority of quadrupeds, which generally have their upper and exposed surfaces more thickly clothed than the lower surface. Nevertheless, the fact that the other members of the order of Primates, to which man belongs, although inhabiting various hot regions, are well clothed with hair, generally thickest on the upper surface,217 is strongly opposed to the supposition that man became naked through the action of the sun. I am inclined to believe, as we shall see under sexual selection, that man, or rather primarily woman, became divested of hair for ornamental purposes; and according to this belief it is not 150surprising that man should differ so greatly in hairiness from all his lower brethren, for characters gained through sexual selection often differ in closely-related forms to an extraordinary degree.

According to a popular impression, the absence of a tail is eminently distinctive of man; but as those apes which come nearest to man are destitute of this organ, its disappearance does not especially concern us. Nevertheless it may be well to own that no explanation, as far as I am aware, has ever been given of the loss of the tail by certain apes and man. Its loss, however, is not surprising, for it sometimes differs remarkably in length in species of the same genera: thus in some species of Macacus the tail is longer than the whole body, consisting of twenty-four vertebræ; in others it consists of a scarcely visible stump, containing only three or four vertebræ. In some kinds of baboons there are twenty-five, whilst in the mandrill there are ten very small stunted caudal vertebræ, or, according to Cuvier,218 sometimes only five. This great diversity in the structure and length of the tail in animals belonging to the same genera, and following nearly the same habits of life, renders it probable that the tail is not of much importance to them; and if so, we might have expected that it would sometimes have become more or less rudimentary, in accordance with what we incessantly see with other structures. The tail almost always tapers towards the end whether it be long or short; and this, I presume, results from the atrophy, through disuse, of the terminal muscles together with their arteries and nerves, leading to the atrophy of the terminal bones. With respect 151to the os coccyx, which in man and the higher apes manifestly consists of the few basal and tapering segments of an ordinary tail, I have heard it asked how could these have become completely embedded within the body; but there is no difficulty in this respect, for in many monkeys the basal segments of the true tail are thus embedded. For instance, Mr. Murie informs me that in the skeleton of a not full-grown Macacus inornatus, he counted nine or ten caudal vertebræ, which altogether were only 1·8 inch in length. Of these the three basal ones appeared to have been embedded; the remainder forming the free part of the tail, which was only one inch in length, and half an inch in diameter. Here, then, the three embedded caudal vertebræ plainly correspond with the four coalesced vertebræ of the human os coccyx.

I have now endeavoured to shew that some of the most distinctive characters of man have in all probability been acquired, either directly, or more commonly indirectly, through natural selection. We should bear in mind that modifications in structure or constitution, which are of no service to an organism in adapting it to its habits of life, to the food which it consumes, or passively to the surrounding conditions, cannot have been thus acquired. We must not, however, be too confident in deciding what modifications are of service to each being: we should remember how little we know about the use of many parts, or what changes in the blood or tissues may serve to fit an organism for a new climate or some new kind of food. Nor must we forget the principle of correlation, by which, as Isidore Geoffroy has shewn in the case of man, many fit-range deviations of structure are tied together. Independently of correlation, a change in one part often leads152 through the increased or decreased use of other parts, to other changes of a quite unexpected nature. It is also well to reflect on such facts, as the wonderful growth of galls on plants caused by the poison of an insect, and on the remarkable changes of colour in the plumage of parrots when fed on certain fishes, or inoculated with the poison of toads;219 for we can thus see that the fluids of the system, if altered for some special purpose, might induce other strange changes. We should especially bear in mind that modifications acquired and continually used during past ages for some useful purpose would probably become firmly fixed and might be long inherited.

Thus a very large yet undefined extension may safely be given to the direct and indirect results of natural selection; but I now admit, after reading the essay by Nägeli on plants, and the remarks by various authors with respect to animals, more especially those recently made by Professor Broca, that in the earlier editions of my 'Origin of Species' I probably attributed too much to the action of natural selection or the survival of the fittest. I have altered the fifth edition of the Origin so as to confine my remarks to adaptive changes of structure. I had not formerly sufficiently considered the existence of many structures which appear to be, as far as we can judge, neither beneficial nor injurious; and this I believe to be one of the greatest oversights as yet detected in my work. I may be permitted to say as some excuse, that I had two distinct objects in view, firstly, to shew that species had not been separately created, and secondly, that natural selection had been the chief agent of change, though largely aided by the 153inherited effects of habit, and slightly by the direct action of the surrounding conditions. Nevertheless I was not able to annul the influence of my former belief, then widely prevalent, that each species had been purposely created; and this led to my tacitly assuming that every detail of structure, excepting rudiments, was of some special, though unrecognised, service. Any one with this assumption in his mind would naturally extend the action of natural selection, either during past or present times, too far. Some of those who admit the principle of evolution, but reject natural selection, seem to forget, when criticising my book, that I had the above two objects in view; hence if I have erred in giving to natural selection great power, which I am far from admitting, or in having exaggerated its power, which is in itself probable, I have at least, as I hope, done good service in aiding to overthrow the dogma of separate creations.

That all organic beings, including man, present many modifications of structure which are of no service to them at present, nor have been formerly, is, as I can now see, probable. We know not what produces the numberless slight differences between the individuals of each species, for reversion only carries the problem a few steps backwards; but each peculiarity must have had its own efficient cause. If these causes, whatever they may be, were to act more uniformly and energetically during a lengthened period (and no reason can be assigned why this should not sometimes occur), the result would probably be not mere slight individual differences, but well-marked, constant modifications. Modifications which are in no way beneficial cannot have been kept uniform through natural selection, though any which were injurious would have been thus eliminated. Uniformity of character would, however,154 naturally follow from, the assumed uniformity of the exciting causes, and likewise from the free intercrossing of many individuals. The same organism might acquire in this manner during successive periods successive modifications, and these would be transmitted in a nearly uniform state as long as the exciting causes remained the same and there was free intercrossing. With respect to the exciting causes we can only say, as when speaking of so-called spontaneous variations, that they relate much more closely to the constitution of the varying organism, than to the nature of the conditions to which it has been subjected.

Conclusion.—In this chapter we have seen that as man at the present day is liable, like every other animal, to multiform individual differences or slight variations, so no doubt were the early progenitors of man; the variations being then as now induced by the same general causes, and governed by the same general and complex laws. As all animals tend to multiply beyond their means of subsistence, so it must have been with the progenitors of man; and this will inevitably have led to a struggle for existence and to natural selection. This latter process will have been greatly aided by the inherited effects of the increased use of parts; these two processes incessantly reacting on each other. It appears, also, as we shall hereafter see, that various unimportant characters have been acquired by man through sexual selection. An unexplained residuum of change, perhaps a large one, must be left to the assumed uniform action of those unknown agencies, which occasionally induce strongly-marked and abrupt deviations of structure in our domestic productions.

Judging from the habits of savages and of the greater number of the Quadrumana, primeval men, and even155 the ape-like progenitors of man, probably lived in society. With strictly social animals, natural selection sometimes acts indirectly on the individual, through the preservation of variations which are beneficial only to the community. A community including a large number of well-endowed individuals increases in number and is victorious over other and less well-endowed communities; although each separate member may gain no advantage over the other members of the same community. With associated insects many remarkable structures, which are of little or no service to the individual or its own offspring, such as the pollen-collecting apparatus, or the sting of the worker-bee, or the great jaws of soldier-ants, have been thus acquired. With the higher social animals, I am not aware that any structure has been modified solely for the good of the community, though some are of secondary service to it. For instance, the horns of ruminants and the great canine teeth of baboons appear to have been acquired by the males as weapons for sexual strife, but they are used in defence of the herd or troop. In regard to certain mental faculties the case, as we shall see in the following chapter, is wholly different; for these faculties have been chiefly, or even exclusively, gained for the benefit of the community; the individuals composing the community being at the same time indirectly benefited.

It has often been objected to such views as the foregoing, that man is one of the most helpless and defenceless creatures in the world; and that during his early and less well-developed condition he would have been still more helpless. The Duke of Argyll, for instance, insists220 that "the human frame has diverged from 156the structure of brutes, in the direction of greater physical helplessness and weakness. That is to say, it is a divergence which of all others it is most impossible to ascribe to mere natural selection." He adduces the naked and unprotected state of the body, the absence of great teeth or claws for defence, the little strength of man, his small speed in running, and his slight power of smell, by which to discover food or to avoid danger. To these deficiencies there might have been added the still more serious loss of the power of quickly climbing trees, so as to escape from enemies. Seeing that the unclothed Fuegians can exist under their wretched climate, the loss of hair would not have been a great injury to primeval man, if he inhabited a warm country. When we compare defenceless man with the apes, many of which are provided with formidable canine teeth, we must remember that these in their fully-developed condition are possessed by the males alone, being chiefly used by them for fighting with their rivals; yet the females which are not thus provided, are able to survive.

In regard to bodily size or strength, we do not know whether man is descended from some comparatively small species, like the chimpanzee, or from one as powerful as the gorilla; and, therefore, we cannot say whether man has become larger and stronger, or smaller and weaker, in comparison with his progenitors. We should, however, bear in mind that an animal possessing great size, strength, and ferocity, and which, like the gorilla, could defend itself from all enemies, would probably, though not necessarily, have failed to become social; and this would most effectually have checked the acquirement by man of his higher mental qualities, such as sympathy and the love of his fellow-creatures. Hence it might have been an immense157 advantage to man to have sprung from some comparatively weak creature.

The slight corporeal strength of man, his little speed, his want of natural weapons, &c., are more than counterbalanced, firstly by his intellectual powers, through which he has, whilst still remaining in a barbarous state, formed for himself weapons, tools, &c., and secondly by his social qualities which lead him to give aid to his fellow-men and to receive it in return. No country in the world abounds in a greater degree with dangerous beasts than Southern

Africa; no country presents more fearful physical hardships than the Arctic regions; yet one of the puniest races, namely, the Bushmen, maintain themselves in Southern Africa, as do the dwarfed Esquimaux in the Arctic regions. The early progenitors of man were, no doubt, inferior in intellect, and probably in social disposition, to the lowest existing savages; but it is quite conceivable that they might have existed, or even flourished, if, whilst they gradually lost their brute-like powers, such as climbing trees, &c., they at the same time advanced in intellect. But granting that the progenitors of man were far more helpless and defenceless than any existing savages, if they had inhabited some warm continent or large island, such as Australia or New Guinea, or Borneo (the latter island being now tenanted by the orang), they would not have been exposed to any special danger. In an area as large as one of these islands, the competition between tribe and tribe would have been sufficient, under favourable conditions, to have raised man, through the survival of the fittest, combined with the inherited effects of habit, to his present high position in the organic scale.

CHAPTER V.

On the Development of the Intellectual and Moral Faculties during Primeval and Civilised Times.

The advancement of the intellectual powers through natural selection—Importance of imitation—Social and moral faculties—Their development within the limits of the same tribe—Natural selection as affecting civilised nations—Evidence that civilised nations were once barbarous.

The subjects to be discussed in this chapter are of the highest interest, but are treated by me in a most imperfect and fragmentary manner. Mr. Wallace, in an admirable paper before referred to,[221] argues that man after he had partially acquired those intellectual and moral faculties which distinguish him from the lower animals, would have been but little liable to have had his bodily structure modified through natural selection or any other means. For man is enabled through his mental faculties "to keep with an unchanged body in harmony with the changing universe." He has great power of adapting his habits to new conditions of life. He invents weapons, tools and various stratagems, by which he procures food and defends himself. When he migrates into a colder climate he uses clothes, builds sheds, and makes fires; and, by the aid of fire, cooks food otherwise indigestible. He aids his fellow-men in many ways, and anticipates future events. Even at a remote period he practised some subdivision of labour.

159

The lower animals, on the other hand, must have their bodily structure modified in order to survive under greatly changed conditions. They must be rendered stronger, or acquire more effective teeth or claws, in order to defend themselves from new enemies; or they must be reduced in size so as to escape detection and danger. When they migrate into a colder climate they must become clothed with thicker fur, or have their constitutions altered. If they fail to be thus modified, they will cease to exist.

The case, however, is widely different, as Mr. Wallace has with justice insisted, in relation to the intellectual and moral faculties of man. These faculties are variable; and we have every reason to believe that the variations tend to be inherited. Therefore, if they were formerly of high importance to primeval man and to his ape-like progenitors, they would have been perfected or advanced through natural selection. Of the high importance of the intellectual faculties there can be no doubt, for man mainly owes to them his preeminent position in the world. We can see that, in the rudest state of society, the individuals who were the most sagacious, who invented and used the best weapons or traps, and who were best able to defend themselves, would rear the greatest number of offspring. The tribes which included the largest number of men thus endowed would increase in number and supplant other tribes. Numbers depend primarily on the means of subsistence, and this, partly on the physical nature of the country, but in a much higher degree on the arts which are there practised. As a tribe increases and is victorious, it is often still further increased by the absorption of other tribes.222 The stature and strength of the men of a tribe 160are likewise of some importance for its success, and these depend in part on the nature and amount of the food which can be obtained. In Europe the men of the Bronze period were supplanted by a more powerful and, judging from their sword-handles, larger-handed race;223 but their success was probably due in a much higher degree to their superiority in the arts.

All that we know about savages, or may infer from their traditions and from old monuments, the history of which is quite forgotten by the present inhabitants, shew that from the remotest times successful tribes have supplanted other tribes. Relics of extinct or forgotten tribes have been discovered throughout the civilised regions of the earth, on the wild plains of America, and on the isolated islands in the Pacific Ocean. At the present day civilised nations are everywhere supplanting barbarous nations, excepting where the climate opposes a deadly barrier; and they succeed mainly, though not exclusively, through their arts, which are the products of the intellect. It is, therefore, highly probable that with mankind the intellectual faculties have been gradually perfected through natural selection; and this conclusion is sufficient for our purpose. Undoubtedly it would have been very interesting to have traced the development of each separate faculty from the state in which it exists in the lower animals to that in which it exists in man; but neither my ability nor knowledge permit the attempt.

It deserves notice that as soon as the progenitors of man became social (and this probably occurred at a very early period), the advancement of the intellectual faculties will have been aided and modified in an important manner, of which we see only traces in 161the lower animals, namely, through the principle of imitation, together with reason and experience. Apes are much given to imitation, as are the lowest savages; and the simple fact previously referred to, that after a time no animal can be caught in the same place by the same sort of trap, shews that animals learn by experience, and imitate each others' caution. Now, if some one man in a tribe, more sagacious than the others, invented a new snare or weapon, or other means of attack or defence, the plainest self-interest, without the assistance of much reasoning power, would prompt the other members to imitate him; and all would thus profit. The habitual practice of each new art must likewise in some slight degree strengthen the intellect. If the new invention were an important one, the tribe would increase in number, spread, and supplant other tribes. In a tribe thus rendered more numerous there would always be a rather better chance of the birth of other superior and inventive members. If such men left children to inherit their mental superiority, the chance of the birth of still more ingenious members would be somewhat better, and in a very small tribe decidedly better. Even if they left no children, the tribe would still include their blood-relations; and it has been ascertained by agriculturists224 that by preserving and breeding from the family of an animal, which when slaughtered was found to be valuable, the desired character has been obtained.

Turning now to the social and moral faculties. In order that primeval men, or the ape-like progenitors of man, should have become social, they must have 162acquired the same instinctive feelings which impel other animals to live in a body; and they no doubt exhibited the same general disposition. They would have felt uneasy when separated from their comrades, for whom they would have felt some degree of love; they would have warned each other of danger, and have given mutual aid in attack or defence. All this implies some degree of sympathy, fidelity, and courage. Such social qualities, the paramount importance of which to the lower animals is disputed by no one, were no doubt acquired by the progenitors of man in a similar manner, namely, through natural selection, aided by inherited habit. When two tribes of primeval man, living in the same country, came into competition, if the one tribe included (other circumstances being equal) a greater number of courageous, sympathetic, and faithful members, who were always ready to warn each other of danger, to aid and defend each other, this tribe would without doubt succeed best and conquer the other. Let it be borne in mind how all-important, in the never-ceasing wars of savages, fidelity and courage must be. The advantage which disciplined soldiers have over undisciplined hordes follows chiefly from the confidence which each man feels in his comrades. Obedience, as Mr. Bagehot has well shewn, 225 is of the highest value, for any form of government is better than none. Selfish and contentious people will not cohere, and without coherence nothing can be effected. A tribe possessing the above qualities in a high degree would spread and be victorious over other tribes; but in the course of time it would, judging from all past history, be in its turn overcome by some other 163and still more highly endowed tribe. Thus the social and moral qualities would tend slowly to advance and be diffused throughout the world.

But it may be asked, how within the limits of the same tribe did a large number of members first become endowed with these social and moral qualities, and how was the standard of excellence raised? It is extremely doubtful whether the offspring of the more sympathetic and benevolent parents, or of those which were the most faithful to their comrades, would be reared in greater number than the children of selfish and treacherous parents of the same tribe. He who was ready to sacrifice his life, as many a savage has been, rather than betray his comrades, would often leave no offspring to inherit his noble nature. The bravest men, who were always willing to come to the front in war, and who freely risked their lives for others, would on an average perish in larger number than other men. Therefore it seems scarcely possible (bearing in mind that we are not here speaking of one tribe being victorious over another) that the number of men gifted with such virtues, or that the standard of their excellence, could be increased through natural selection, that is, by the survival of the fittest.

Although the circumstances which lead to an increase in the number of men thus endowed within the same tribe are too complex to be clearly followed out, we can trace some of the probable steps. In the first place, as the reasoning powers and foresight of the members became improved, each man would soon learn from experience that if he aided his fellow-men, he would commonly receive aid in return. From this low motive he might acquire the habit of aiding his fellows; and the habit of performing benevolent actions certainly strengthens the feeling of sympathy, which gives the164 first impulse to benevolent actions. Habits, moreover, followed during many generations probably tend to be inherited.

But there is another and much more powerful stimulus to the development of the social virtues, namely, the praise and the blame of our fellow-men. The love of approbation and the dread of infamy, as well as the bestowal of praise or blame, are primarily due, as we have seen in the third chapter, to the instinct of sympathy; and this instinct no doubt was originally acquired, like all the other social instincts, through natural selection. At how early a period the progenitors of man, in the course of their development, became capable of feeling and being impelled by the praise or blame of their fellow-creatures, we cannot, of course, say. But it appears that even dogs appreciate encouragement, praise, and blame. The rudest savages feel the sentiment of glory, as they clearly show by preserving the trophies of their prowess, by their habit of excessive boasting, and even by the extreme care which they take of their personal appearance and decorations; for unless they regarded the opinion of their comrades, such habits would be senseless.

They certainly feel shame at the breach of some of their lesser rules; but how far they experience remorse is doubtful. I was at first surprised that I could not recollect any recorded instances of this feeling in savages; and Sir J. Lubbock226 states that he knows of none. But if we banish from our minds all cases given in novels and plays and in death-bed confessions made to priests, I doubt whether many of us have actually witnessed remorse; though we may have often seen shame and contrition for smaller offences. Remorse is 165a deeply hidden feeling. It is incredible that a savage, who will sacrifice his life rather than betray his tribe, or one who will deliver himself up as a prisoner rather than break his parole,227 would not feel remorse in his inmost soul, though he might conceal it, if he had failed in a duty which he held sacred.

We may therefore conclude that primeval man, at a very remote period, would have been influenced by the praise and blame of his fellows. It is obvious, that the members of the same tribe would approve of conduct which appeared to them to be for the general good, and would reprobate that which appeared evil. To do good unto others—to do unto others as ye would they should do unto you,—is the foundation-stone of morality. It is, therefore, hardly possible to exaggerate the importance during rude times of the love of praise and the dread of blame. A man who was not impelled by any deep, instinctive feeling, to sacrifice his life for the good of others, yet was roused to such actions by a sense of glory, would by his example excite the same wish for glory in other men, and would strengthen by exercise the noble feeling of admiration. He might thus do far more good to his tribe than by begetting offspring with a tendency to inherit his own high character.

With increased experience and reason, man perceives the more remote consequences of his actions, and the self-regarding virtues, such as temperance, chastity, &c., which during early times are, as we have before seen, utterly disregarded, come to be highly esteemed or even held sacred. I need not, however, repeat what I have said on this head in the third chapter. Ultimately a highly complex sentiment, having its first origin in the 166social instincts, largely guided by the approbation of our fellow-men, ruled by reason, self-interest, and in later times by deep religious feelings, confirmed by instruction and habit, all combined, constitute our moral sense or conscience.

It must not be forgotten that although a high standard of morality gives but a slight or no advantage to each individual man and his children over the other men of the same tribe, yet that an advancement in the standard of morality and an increase in the number of well-endowed men will certainly give an immense advantage to one tribe over another. There can be no doubt that a tribe including many members who, from possessing in a high degree the spirit of patriotism, fidelity, obedience, courage, and sympathy, were always ready to give aid to each other and to sacrifice themselves for the common good, would be victorious over most other tribes; and this would be natural selection. At all times throughout the world tribes have supplanted other tribes; and as morality is one element in their success, the standard of morality and the number of well-endowed men will thus everywhere tend to rise and increase.

It is, however, very difficult to form any judgment why one particular tribe and not another has been successful and has risen in the scale of civilisation. Many savages are in the same condition as when first discovered several centuries ago. As Mr. Bagehot has remarked, we are apt to look at progress as the normal rule in human society; but history refutes this. The ancients did not even entertain the idea; nor do the oriental nations at the present day. According to another high authority, Mr. Maine,228 "the greatest part of mankind has never 167shewn a particle of desire that its civil institutions should be improved." Progress seems to depend on many concurrent favourable conditions, far too complex to be followed out. But it has often been remarked, that a cool climate from leading to industry and the various arts has been highly favourable, or even indispensable for this end. The Esquimaux, pressed by hard necessity, have succeeded in many ingenious inventions, but their climate has been too severe for continued progress. Nomadic habits, whether over wide plains, or through the dense forests of the tropics, or along the shores of the sea, have in every case been highly detrimental. Whilst observing the barbarous inhabitants of Tierra del Fuego, it struck me that the possession of some property, a fixed abode, and the union of many families under a chief, were the indispensable requisites for civilisation. Such habits almost necessitate the cultivation of the ground; and the first steps in cultivation would probably result, as I have elsewhere shewn,229 from some such accident as the seeds of a fruit-tree falling on a heap of refuse and producing an unusually fine variety. The problem, however, of the first advance of savages towards civilisation is at present much too difficult to be solved.

Natural Selection as affecting Civilised Nations.—In the last and present chapters I have considered the advancement of man from a former semi-human condition to his present state as a barbarian. But some remarks on the agency of natural selection on civilised nations may be here worth adding. This subject has been ably discussed by Mr. W. R. Greg,230 and previously 168by Mr. Wallace and Mr. Galton.231 Most of my remarks are taken from these three authors. With savages, the weak in body or mind are soon eliminated; and those that survive commonly exhibit a vigorous state of health. We civilised men, on the other hand, do our utmost to check the process of elimination; we build asylums for the imbecile, the maimed, and the sick; we institute poor-laws; and our medical men exert their utmost skill to save the life of every one to the last moment. There is reason to believe that vaccination has preserved thousands, who from a weak constitution would formerly have succumbed to small-pox. Thus the weak members of civilised societies propagate their kind. No one who has attended to the breeding of domestic animals will doubt that this must be highly injurious to the race of man. It is surprising how soon a want of care, or care wrongly directed, leads to the degeneration of a domestic race; but excepting in the case of man himself, hardly any one is so ignorant as to allow his worst animals to breed.

The aid which we feel impelled to give to the helpless is mainly an incidental result of the instinct of sympathy, which was originally acquired as part of the social instincts, but subsequently rendered, in the manner previously indicated, more tender and more widely diffused. Nor could we check our sympathy, if so urged by hard reason, without deterioration in the 169noblest part of our nature. The surgeon may harden himself whilst performing an operation, for he knows that he is acting for the good of his patient; but if we were intentionally to neglect the weak and helpless, it could only be for a contingent benefit, with a certain and great present evil. Hence we must bear without complaining the undoubtedly bad effects of the weak surviving and propagating their kind; but there appears to be at least one check in steady action, namely the weaker and inferior members of society not marrying so freely as the sound; and this check might be indefinitely increased, though this is more to be hoped for than expected, by the weak in body or mind refraining from marriage.

In all civilised countries man accumulates property and bequeaths it to his children. So that the children in the same country do not by any means start fair in the race for success. But this is far from an unmixed evil; for without the accumulation of capital the arts could not progress; and it is chiefly through their power that the civilised races have extended, and are now everywhere extending, their range, so as to take the place of the lower races. Nor does the moderate accumulation of wealth interfere with the process of selection. When a poor man becomes rich, his children enter trades or professions in which there is struggle enough, so that the able in body and mind succeed best. The presence of a body of well-instructed men, who have not to labour for their daily bread, is important to a degree which cannot be over-estimated; as all high intellectual work is carried on by them, and on such work material progress of all kinds mainly depends, not to mention other and higher advantages. No doubt wealth when very great tends to convert men into useless drones, but their number is never large; and some degree of elimi170nation here occurs, as we daily see rich men, who happen to be fools or profligate, squandering away all their wealth.

Primogeniture with entailed estates is a more direct evil, though it may formerly have been a great advantage by the creation of a dominant class, and any government is better than anarchy. The eldest sons, though they may be weak in body or mind, generally marry, whilst the younger sons, however superior in these respects, do not so generally marry. Nor can worthless eldest sons with entailed estates squander their wealth. But here, as elsewhere, the relations of civilised life are so complex that some compensatory checks intervene. The men who are rich through primogeniture are able to select generation after generation the more beautiful and charming women; and these must generally be healthy in body and active in mind. The evil consequences, such as they may be, of the continued preservation of the same line of descent, without any selection, are checked by men of rank always wishing to increase their wealth and power; and this they effect by marrying heiresses. But the daughters of parents who have produced single children, are themselves, as Mr. Galton has shewn,232 apt to be sterile; and thus noble families are continually cut off in the direct line, and their wealth flows into some side channel; but unfortunately this channel is not determined by superiority of any kind.

Although civilisation thus checks in many ways the action of natural selection, it apparently favours, by means of improved food and the freedom from occasional hardships, the better development of the body. This may be inferred from civilised men having been 171found, wherever compared, to be physically stronger than savages. They appear also to have equal powers of endurance, as has been proved in many adventurous expeditions. Even the great luxury of the rich can be but little detrimental; for the expectation of life of our aristocracy, at all ages and of both sexes, is very little inferior to that of healthy English lives in the lower classes.233

We will now look to the intellectual faculties alone. If in each grade of society the members were divided into two equal bodies, the one including the intellectually superior and the other the inferior, there can be little doubt that the former would succeed best in all occupations and rear a greater number of children. Even in the lowest walks of life, skill and ability must be of some advantage, though in many occupations, owing to the great division of labour, a very small one. Hence in civilised nations there will be some tendency to an increase both in the number and in the standard of the intellectually able. But I do not wish to assert that this tendency may not be more than counterbalanced in other ways, as by the multiplication of the reckless and improvident; but even to such as these, ability must be some advantage.

It has often been objected to views like the foregoing, that the most eminent men who have ever lived have left no offspring to inherit their great intellect. Mr. Galton says,[234] "I regret I am unable to solve the simple question whether, and how far, men and women who are prodigies of genius are infertile. I have, however, shewn that men of eminence are by no means so."

172

Great lawgivers, the founders of beneficent religions, great philosophers and discoverers in science, aid the progress of mankind in a far higher degree by their works than by leaving a numerous progeny. In the case of corporeal structures, it is the selection of the slightly better-endowed and the elimination of the slightly less well-endowed individuals, and not the preservation of strongly-marked and rare anomalies, that leads to the advancement of a species.[235] So it will be with the intellectual faculties, namely from the somewhat more able men in each grade of society succeeding rather better than the less able, and consequently increasing in number, if not otherwise prevented. When in any nation the standard of intellect and the number of intellectual men have increased, we may expect from the law of the deviation from an average, as shewn by Mr. Galton, that prodigies of genius will appear somewhat more frequently than before.

In regard to the moral qualities, some elimination of the worst dispositions is always in progress even in the most civilised nations. Malefactors are executed, or imprisoned for long periods, so that they cannot freely transmit their bad qualities. Melancholic and insane persons are confined, or commit suicide. Violent and quarrelsome men often come to a bloody end. Restless men who will not follow any steady occupation—and this relic of barbarism is a great check to civilisation[236]—emigrate to newly-settled countries, where they prove useful pioneers. Intemperance is so highly destructive, that the expectation of life of the intemperate, at the age, for instance, of thirty, is only 13.8 years; whilst for the rural labourers of England at the same age it is 17340·59 years.[237] Profligate women bear few children, and profligate men rarely marry; both suffer from disease. In the breeding of domestic animals, the elimination of those individuals, though few in number, which are in any marked manner inferior, is by no means an unimportant element towards success. This especially holds good with injurious characters which tend to reappear through reversion, such as blackness in sheep; and with mankind some of the worst dispositions, which occasionally without any assignable cause make their appearance in families, may perhaps be reversions to a savage state, from which we are not removed by very many generations. This view seems indeed recognised in the common expression that such men are the black sheep of the family.

With civilised nations, as far as an advanced standard of morality, and an increased number of fairly well-endowed men are concerned, natural selection apparently effects but little; though the fundamental social instincts were originally thus gained. But I have already said enough, whilst treating of the lower races, on the causes which lead to the advance of morality, namely, the approbation of our fellow-men—the strengthening of our sympathies by habit—example and imitation—reason—experience and even self-interest—instruction during youth, and religious feelings.

A most important obstacle in civilised countries to an increase in the number of men of a superior class has been strongly urged by Mr. Greg and Mr. Galton,[238] 174namely, the fact that the very poor and reckless, who are often degraded by vice, almost invariably marry early, whilst the careful and frugal, who are generally otherwise virtuous, marry late in life, so that they may be able to support themselves and their children in comfort. Those who marry early produce within a given period not only a greater number of generations, but, as shewn by Dr. Duncan,239 they produce many more children. The children, moreover, that are born by mothers during the prime of life are heavier and larger, and therefore probably more vigorous, than those born at other periods. Thus the reckless, degraded, and often vicious members of society, tend to increase at a quicker rate than the provident and generally virtuous members. Or as Mr. Greg puts the case: "The careless, squalid, unaspiring Irishman multiplies like rabbits: the frugal, foreseeing, self-respecting, ambitious Scot, stern in his morality, spiritual in his faith, sagacious and disciplined in his intelligence, passes his best years in struggle and in celibacy, marries late, and leaves few behind him. Given a land originally peopled by a thousand Saxons and a thousand Celts—and in a dozen generations five-sixths of the population would be Celts, but five-sixths of the property, of the power, of the intellect, would belong to the one-sixth of Saxons that remained. In the eternal 'struggle for existence,' it would be the inferior and less favoured race that had prevailed—and prevailed by virtue not of its good qualities but of its faults."

There are, however, some checks to this downward tendency. We have seen that the intemperate suffer 175from a high rate of mortality, and the extremely profligate leave few offspring. The poorest classes crowd into towns, and it has been proved by Dr. Stark from the statistics of ten years in Scotland,240 that at all ages the death-rate is higher in towns than in rural districts, "and during the first five years of life the town death-rate is almost exactly double that of the rural districts." As these returns include both the rich and the poor, no doubt more than double the number of births would be requisite to keep up the number of the very poor inhabitants in the towns, relatively to those in the country. With women, marriage at too early an age is highly injurious; for it has been found in France that, "twice as many wives under twenty die in the year, as died out of the same number of the unmarried." The mortality, also, of husbands under twenty is "excessively high,"241 but what the cause of this may be seems doubtful. Lastly, if the men who prudently delay marrying until they can bring up their families in comfort, were to select, as they often do, women in the prime of life, the rate of increase in the better class would be only slightly lessened.

It was established from an enormous body of statistics, taken during 1853, that the unmarried men throughout France, between the ages of twenty and eighty, die in a much larger proportion than the married: for instance, out of every 1000 unmarried men, between the ages of twenty and thirty, 11·3 annually died, whilst of the married only 6·5 died.242 A similar law was proved to 176hold good, during the years 1863 and 1864, with the entire population above the age of twenty in Scotland: for instance, out of every 1000 unmarried men, between the ages of twenty and thirty, 14·97 annually died, whilst of the married only 7·24 died, that is less than half.243 Dr. Stark remarks on this, "Bachelorhood is more destructive to life than the most unwholesome trades, or than residence in an unwholesome house or district where there has never been the most distant attempt at sanitary improvement." He considers that the lessened mortality is the direct result of "marriage, and the more regular domestic habits which attend that state." He admits, however, that the intemperate, profligate, and criminal classes, whose duration of life is low, do not commonly marry; and it must likewise be admitted that men with a weak constitution, ill health, or any great infirmity in body or mind, will often not wish to marry, or will be rejected. Dr. Stark seems to have come to the conclusion that marriage in itself is a main cause of prolonged life, from finding that aged married men still have a considerable advantage in this respect over the unmarried of the same advanced age; but every one must have known instances of men, who with weak health during youth did not marry, and yet have survived to old age, though remaining weak and therefore always with a lessened chance of life. There is another remarkable circumstance which seems to support Dr. Stark's conclusion, namely, that widows and widowers in France suffer in comparison with the married a very heavy rate of mortality; but Dr. Farr attributes this to the poverty and 177evil habits consequent on the disruption of the family, and to grief. On the whole we may conclude with Dr. Farr that the lesser mortality of married than of unmarried men, which seems to be a general law, "is mainly due to the constant elimination of imperfect types, and to the skilful selection of the finest individuals out of each successive generation;" the selection relating only to the marriage state, and acting on all corporeal, intellectual, and moral qualities. We may, therefore, infer that sound and good men who out of prudence remain for a time unmarried do not suffer a high rate of mortality.

If the various checks specified in the two last paragraphs, and perhaps others as yet unknown, do not prevent the reckless, the vicious and otherwise inferior members of society from increasing at a quicker rate than the better class of men, the nation will retrograde, as has occurred too often in the history of the world. We must remember that progress is no invariable rule. It is most difficult to say why one civilised nation rises, becomes more powerful, and spreads more widely, than another; or why the same nation progresses more at one time than at another. We can only say that it depends on an increase in the actual number of the population, on the number of the men endowed with high intellectual and moral faculties, as well as on their standard of excellence. Corporeal structure, except so far as vigour of body leads to vigour of mind, appears to have little influence.

It has been urged by several writers that as high intellectual powers are advantageous to a nation, the old Greeks, who stood some grades higher in intellect than any race that has ever existed,244 ought to have 178risen, if the power of natural selection were real, still higher in the scale, increased in number, and stocked the whole of Europe. Here we have the tacit assumption, so often made with respect to corporeal structures, that there is some innate tendency towards continued development in mind and body. But development of all kinds depends on many concurrent favourable circumstances. Natural selection acts only in a tentative manner. Individuals and races may have acquired certain indisputable advantages, and yet have perished from failing in other characters. The Greeks may have retrograded from a want of coherence between the many small states, from the small size of their whole country, from the practice of slavery, or from extreme sensuality; for they did not succumb until "they were enervated and corrupt to the very core."245 The western nations of Europe, who now so immeasurably surpass their former savage progenitors and stand at the summit of civilisation, owe little or none of their superiority to direct inheritance from the old Greeks; though they owe much to the written works of this wonderful people.

Who can positively say why the Spanish nation, so dominant at one time, has been distanced in the race. The awakening of the nations of Europe from the dark ages is a still more perplexing problem. At this early period, as Mr. Galton246 has remarked, almost all the men of a gentle nature, those given to meditation or culture of the mind, had no refuge except in the bosom of the Church which demanded celibacy; 179and this could hardly fail to have had a deteriorating influence on each successive generation. During this same period the Holy Inquisition selected with extreme care the freest and boldest men in order to burn or imprison them. In Spain alone some of the best men—those who doubted and questioned, and without doubting there can be no progress—were eliminated during three centuries at the rate of a thousand a year. The evil which the Catholic Church has thus effected, though no doubt counterbalanced to a certain, perhaps large extent in other ways, is incalculable; nevertheless, Europe has progressed at an unparalleled rate.

The remarkable success of the English as colonists over other European nations, which is well illustrated by comparing the progress of the Canadians of English and French extraction, has been ascribed to their "daring and persistent energy;" but who can say how the English gained their energy. There is apparently much truth in the belief that the wonderful progress of the United States, as well as the character of the people, are the results of natural selection; the more energetic, restless, and courageous men from all parts of Europe having emigrated during the last ten or twelve generations to that great country, and having there succeeded best.247 Looking to the distant future, I do not think that the Rev. Mr. Zincke takes an exaggerated view when he says:248 "All other series of events—as that which resulted in the culture of mind in Greece, and that which resulted in the empire of Rome—only appear to have purpose and value when viewed in connection with, or rather as subsidiary to ... the great stream of Anglo-Saxon emigration to the west."
180

Obscure as is the problem of the advance of civilisation, we can at least see that a nation which produced during a lengthened period the greatest number of highly intellectual, energetic, brave, patriotic, and benevolent men, would generally prevail over less favoured nations.

Natural selection follows from the struggle for existence; and this from a rapid rate of increase. It is impossible not bitterly to regret, but whether wisely is another question, the rate at which man tends to increase; for this leads in barbarous tribes to infanticide and many other evils, and in civilised nations to abject poverty, celibacy, and to the late marriages of the prudent. But as man suffers from the same physical evils with the lower animals, he has no right to expect an immunity from the evils consequent on the struggle for existence. Had he not been subjected to natural selection, assuredly he would never have attained to the rank of manhood. When we see in many parts of the world enormous areas of the most fertile land peopled by a few wandering savages, but which are capable of supporting numerous happy homes, it might be argued that the struggle for existence had not been sufficiently severe to force man upwards to his highest standard. Judging from all that we know of man and the lower animals, there has always been sufficient variability in the intellectual and moral faculties, for their steady advancement through natural selection. No doubt such advancement demands many favourable concurrent circumstances; but it may well be doubted whether the most favourable would have sufficed, had not the rate of increase been rapid, and the consequent struggle for existence severe to an extreme degree.

On the evidence that all civilised nations were once barbarous.—As we have had to consider the steps by which181 some semi-human creature has been gradually raised to the rank of man in his most perfect state, the present subject cannot be quite passed over. But it has been treated in so full and admirable a manner by Sir J. Lubbock,249 Mr. Tylor, Mr. M'Lennan, and others, that I need here give only the briefest summary of their results. The arguments recently advanced by the Duke of Argyll250 and formerly by Archbishop Whately, in favour of the belief that man came into the world as a civilised being and that all savages have since undergone degradation, seem to me weak in comparison with those advanced on the other side. Many nations, no doubt, have fallen away in civilisation, and some may have lapsed into utter barbarism, though on this latter head I have not met with any evidence. The Fuegians were probably compelled by other conquering hordes to settle in their inhospitable country, and they may have become in consequence somewhat more degraded; but it would be difficult to prove that they have fallen much below the Botocudos who inhabit the finest parts of Brazil.

The evidence that all civilised nations are the descendants of barbarians, consists, on the one side, of clear traces of their former low condition in still-existing customs, beliefs, language, &c.; and on the other side, of proofs that savages are independently able to raise themselves a few steps in the scale of civilisation, and have actually thus risen. The evidence on the first head is extremely curious, but cannot be here given: I refer to such cases as that, for instance, of the art of enumeration, which, as Mr. Tylor clearly shews by the words still used in some places, originated in counting 182the fingers, first of one hand and then of the other, and lastly of the toes. We have traces of this in our own decimal system, and in the Roman numerals, which after reaching to the number V., change into VI., &c., when the other hand no doubt was used. So again, "when we speak of three-score and ten, we are counting by the vigesimal system, each score thus ideally made, standing for 20—for 'one man' as a Mexican or Carib would put it."251 According to a large and increasing school of philologists, every language bears the marks of its slow and gradual evolution. So it is with the art of writing, as letters are rudiments of pictorial representations. It is hardly possible to read Mr. M'Lennan's work252 and not admit that almost all civilised nations still retain some traces of such rude habits as the forcible capture of wives. What ancient nation, as the same author asks, can be named that was originally monogamous? The primitive idea of justice, as shewn by the law of battle and other customs of which traces still remain, was likewise most rude. Many existing superstitions are the remnants of former false religious beliefs. The highest form of religion—the grand idea of God hating sin and loving righteousness—was unknown during primeval times.

Turning to the other kind of evidence: Sir J. Lubbock has shewn that some savages have recently improved a little in some of their simpler arts. From the 183extremely curious account which he gives of the weapons, tools, and arts, used or practised by savages in various parts of the world, it cannot be doubted that these have nearly all been independent discoveries, excepting perhaps the art of making fire.253 The Australian boomerang is a good instance of one such independent discovery. The Tahitians when first visited had advanced in many respects beyond the inhabitants of most of the other Polynesian islands. There are no just grounds for the belief that the high culture of the native Peruvians and Mexicans was derived from any foreign source;254 many native plants were there cultivated, and a few native animals domesticated. We should bear in mind that a wandering crew from some semi-civilised land, if washed to the shores of America, would not, judging from the small influence of most missionaries, have produced any marked effect on the natives, unless they had already become somewhat advanced. Looking to a very remote period in the history of the world, we find, to use Sir J. Lubbock's well-known terms, a paleolithic and neolithic period; and no one will pretend that the art of grinding rough flint tools was a borrowed one. In all parts of Europe, as far east as Greece, in Palestine, India, Japan, New Zealand, and Africa, including Egypt, flint tools have been discovered in abundance; and of their use the existing inhabitants retain no tradition. There is also indirect evidence of their former use by the Chinese and ancient Jews. Hence there can hardly be a doubt that the inhabitants of these many countries, which include nearly the whole civilised world, were once in a barbarous condition. To believe that man was aboriginally 184civilised and then suffered utter degradation in so many regions, is to take a pitiably low view of human nature. It is apparently a truer and more cheerful view that progress has been much more general than retrogression; that man has risen, though by slow and interrupted steps, from a lowly condition to the highest standard as yet attained by him in knowledge, morals, and religion.

CHAPTER VI.
On the Affinities and Genealogy of Man.

Position of man in the animal series—The natural system genealogical—Adaptive characters of slight value—Various small points of resemblance between man and the Quadrumana—Rank of man in the natural system—Birthplace and antiquity of man—Absence of fossil connecting-links—Lower stages in the genealogy of man, as inferred, firstly from his affinities and secondly from his structure—Early androgynous condition of the Vertebrata—Conclusion.

Even if it be granted that the difference between man and his nearest allies is as great in corporeal structure as some naturalists maintain, and although we must grant that the difference between them is immense in mental power, yet the facts given in the previous chapters declare, as it appears to me, in the plainest manner, that man is descended from some lower form, notwithstanding that connecting-links have not hitherto been discovered.

Man is liable to numerous, slight, and diversified variations, which are induced by the same general causes, are governed and transmitted in accordance with the same general laws, as in the lower animals. Man tends to multiply at so rapid a rate that his offspring are necessarily exposed to a struggle for existence, and consequently to natural selection. He has given rise to many races, some of which are so different that they have often been ranked by naturalists as distinct species. His body is constructed on the same homological plan as that of other mammals, independently of the uses to which the several parts may be put. He186 passes through the same phases of embryological development. He retains many rudimentary and useless structures, which no doubt were once serviceable. Characters occasionally make their reappearance in him, which we have every reason to believe were possessed by his early progenitors. If the origin of man had been wholly different from that of all other animals, these various appearances would be mere empty deceptions; but such an admission is incredible. These appearances, on the other hand, are intelligible, at least to a large extent, if man is the co-descendant with other mammals of some unknown and lower form.

Some naturalists, from being deeply impressed with the mental and spiritual powers of man, have divided the whole organic world into three kingdoms, the Human, the Animal, and the Vegetable, thus giving to man a separate kingdom.255 Spiritual powers cannot be compared or classed by the naturalist; but he may endeavour to shew, as I have done, that the mental faculties of man and the lower animals do not differ in kind, although immensely in degree. A difference in degree, however great, does not justify us in placing man in a distinct kingdom, as will perhaps be best illustrated by comparing the mental powers of two insects, namely, a coccus or scale-insect and an ant, which undoubtedly belong to the same class. The difference is here greater, though of a somewhat different kind, than that between man and the highest mammal. The female coccus, whilst young, attaches itself by its proboscis to a plant; sucks the sap but never moves again; is fertilised and lays eggs; and this is its whole history. On the other hand, to describe the habits and mental 187powers of a female ant, would require, as Pierre Huber has shewn, a large volume; I may, however, briefly specify a few points. Ants communicate information to each other, and several unite for the same work, or games of play. They recognise their fellow-ants after months of absence. They build great edifices, keep them clean, close the doors in the evening, and post sentries. They make roads, and even tunnels under rivers. They collect food for the community, and when an object, too large for entrance, is brought to the nest, they enlarge the door, and afterwards build it up again.256 They go out to battle in regular bands, and freely sacrifice their lives for the common weal. They emigrate in accordance with a preconcerted plan. They capture slaves. They keep Aphides as milch-cows. They move the eggs of their aphides, as well as their own eggs and cocoons, into warm parts of the nest, in order that they may be quickly hatched; and endless similar facts could be given. On the whole, the difference in mental power between an ant and a coccus is immense; yet no one has ever dreamed of placing them in distinct classes, much less in distinct kingdoms. No doubt this interval is bridged over by the intermediate mental powers of many other insects; and this is not the case with man and the higher apes. But we have every reason to believe that breaks in the series are simply the result of many forms having become extinct.

Professor Owen, relying chiefly on the structure of the brain, has divided the mammalian series into four sub-classes. One of these he devotes to man; in another he places both the marsupials and the monotremata; so that he makes man as distinct from all other mam188mals as are these two latter groups conjoined. This view has not been accepted, as far as I am aware, by any naturalist capable of forming an independent judgment, and therefore need not here be further considered.

We can understand why a classification founded on any single character or organ—even an organ so wonderfully complex and important as the brain—or on the high development of the mental faculties, is almost sure to prove unsatisfactory. This principle has indeed been tried with hymenopterous insects; but when thus classed by their habits or instincts, the arrangement proved thoroughly artificial.257 Classifications may, of course, be based on any character whatever, as on size, colour, or the element inhabited; but naturalists have long felt a profound conviction that there is a natural system. This system, it is now generally admitted, must be, as far as possible, genealogical in arrangement,—that is, the co-descendants of the same form must be kept together in one group, separate from the co-descendants of any other form; but if the parent-forms are related, so will be their descendants, and the two groups together will form a larger group. The amount of difference between the several groups—that is the amount of modification which each has undergone—will be expressed by such terms as genera, families, orders, and classes. As we have no record of the lines of descent, these lines can be discovered only by observing the degrees of resemblance between the beings which are to be classed. For this object numerous points of resemblance are of much more importance than the amount of similarity or dissimilarity in a few points. If two languages were found to resemble each other in a multitude of 189words and points of construction, they would be universally recognised as having sprung from a common source, notwithstanding that they differed greatly in some few words or points of construction. But with organic beings the points of resemblance must not consist of adaptations to similar habits of life: two animals may, for instance, have had their whole frames modified for living in the water, and yet they will not be brought any nearer to each other in the natural system. Hence we can see how it is that resemblances in unimportant structures, in useless and rudimentary organs, and in parts not as yet fully developed or functionally active, are by far the most serviceable for classification;

for they can hardly be due to adaptations within a late period; and thus they reveal the old lines of descent or of true affinity.

We can further see why a great amount of modification in some one character ought not to lead us to separate widely any two organisms. A part which already differs much from the same part in other allied forms has already, according to the theory of evolution, varied much; consequently it would (as long as the organism remained exposed to the same exciting conditions) be liable to further variations of the same kind; and these, if beneficial, would be preserved, and thus continually augmented. In many cases the continued development of a part, for instance, of the beak of a bird, or of the teeth of a mammal, would not be advantageous to the species for gaining its food, or for any other object; but with man we can see no definite limit, as far as advantage is concerned, to the continued development of the brain and mental faculties. Therefore in determining the position of man in the natural or genealogical system, the extreme development of his brain ought not to outweigh a multitude of resem190blances in other less important or quite unimportant points.

The greater number of naturalists who have taken into consideration the whole structure of man, including his mental faculties, have followed Blumenbach and Cuvier, and have placed man in a separate Order, under the title of the Bimana, and therefore on an equality with the Orders of the Quadrumana, Carnivora, &c. Recently many of our best naturalists have recurred to the view first propounded by Linnæus, so remarkable for his sagacity, and have placed man in the same Order with the Quadrumana, under the title of the Primates. The justice of this conclusion will be admitted if, in the first place, we bear in mind the remarks just made on the comparatively small importance for classification of the great development of the brain in man; bearing, also, in mind that the strongly-marked differences between the skulls of man and the Quadrumana (lately insisted upon by Bischoff, Aeby, and others) apparently follow from their differently developed brains. In the second place, we must remember that nearly all the other and more important differences between man and the Quadrumana are manifestly adaptive in their nature, and relate chiefly to the erect position of man; such as the structure of his hand, foot, and pelvis, the curvature of his spine, and the position of his head. The family of seals offers a good illustration of the small importance of adaptive characters for classification. These animals differ from all other Carnivora in the form of their bodies and in the structure of their limbs, far more than does man from the higher apes; yet in every system, from that of Cuvier to the most recent one by Mr. Flower,[258] seals are ranked as a mere family 191in the Order of the Carnivora. If man had not been his own classifier, he would never have thought of founding a separate order for his own reception.

It would be beyond my limits, and quite beyond my knowledge, even to name the innumerable points of structure in which man agrees with the other Primates. Our great anatomist and philosopher, Prof. Huxley, has fully discussed this subject,[259] and has come to the conclusion that man in all parts of his organisation differs less from the higher apes, than these do from the lower members of the same group. Consequently there "is no justification for placing man in a distinct order."

In an early part of this volume I brought forward various facts, shewing how closely man agrees in constitution with the higher mammals; and this agreement, no doubt, depends on our close similarity in minute structure and chemical composition. I gave, as instances, our liability to the same diseases, and to the attacks of allied parasites; our tastes in common for the same stimulants, and the similar effects thus produced, as well as by various drugs; and other such facts.

As small unimportant points of resemblance between man and the higher apes are not commonly noticed in systematic works, and as, when numerous, they clearly reveal our relationship, I will specify a few such points. The relative position of the features are manifestly the same in man and the Quadrumana; and the various emotions are displayed by nearly similar movements of the muscles and skin, chiefly above the eyebrows and round the mouth. Some few expressions are, indeed, almost the same, as in the weeping of certain kinds of monkeys, and in the laughing noise made by others, during which the corners of the mouth are drawn back192wards, and the lower eyelids wrinkled. The external ears are curiously alike. In man the nose is much more prominent than in most monkeys; but we may trace the commencement of an aquiline curvature in the nose of the Hoolock Gibbon; and this in the Semnopithecus nasica is carried to a ridiculous extreme.

The faces of many monkeys are ornamented with beards, whiskers, or moustaches. The hair on the head grows to a great length in some species of Semnopithecus;260 and in the Bonnet monkey (Macacus radiatus) it radiates from a point on the crown, with a parting down the middle, as in man. It is commonly said that the forehead gives to man his noble and intellectual appearance; but the thick hair on the head of the Bonnet monkey terminates abruptly downwards, and is succeeded by such short and fine hair, or down, that at a little distance the forehead, with the exception of the eyebrows, appears quite naked. It has been erroneously asserted that eyebrows are not present in any monkey. In the species just named the degree of nakedness of the forehead differs in different individuals; and Eschricht states261 that in our children the limit between the hairy scalp and the naked forehead is sometimes not well defined; so that here we seem to have a trifling case of reversion to a progenitor, in whom the forehead had not as yet become quite naked.

It is well known that the hair on our arms tends to converge from above and below to a point at the elbow. This curious arrangement, so unlike that in most of the lower mammals, is common to the gorilla, chimpanzee, orang, some species of Hylobates, and even to some few American monkeys. But in Hylobates agilis the hair 193on the fore-arm is directed downwards or towards the wrist in the ordinary manner; and in H. lar it is nearly erect, with only a very slight forward inclination; so that in this latter species it is in a transitional state. It can hardly be doubted that with most mammals the thickness of the hair and its direction on the back is adapted to throw off the rain; even the transverse hairs on the fore-legs of a dog may serve for this end when he is coiled up asleep. Mr. Wallace remarks that the convergence of the hair towards the elbow on the arms of the orang (whose habits he has so carefully studied) serves to throw off the rain, when, as is the custom of this animal, the arms are bent, with the hands clasped round a branch or over its own head. We should, however, bear in mind that the attitude of an animal may perhaps be in part determined by the direction of the hair; and not the direction of the hair by the attitude. If the above explanation is correct in the case of the orang, the hair on our fore-arms offers a curious record of our former state; for no one supposes that it is now of any use in throwing off the rain, nor in our present erect condition is it properly directed for this purpose.

It would, however, be rash to trust too much to the principle of adaptation in regard to the direction of the hair in man or his early progenitors; for it is impossible to study the figures given by Eschricht of the arrangement of the hair on the human fœtus (this being the same as in the adult) and not agree with this excellent observer that other and more complex causes have intervened. The points of convergence seem to stand in some relation to those points in the embryo which are last closed in during development. There appears, also, to exist some relation between the arrangement194 of the hair on the limbs, and the course of the medullary arteries.262

It must not be supposed that the resemblances between man and certain apes in the above and many other points—such as in having a naked forehead, long tresses on the head, &c.—are all necessarily the result of unbroken inheritance from a common progenitor thus characterised, or of subsequent reversion. Many of these resemblances are more probably due to analogous variation, which follows, as I have elsewhere attempted to shew,263 from co-descended organisms having a similar constitution and having been acted on by similar causes inducing variability. With respect to the similar direction of the hair on the fore-arms of man and certain monkeys, as this character is common to almost all the anthropomorphous apes, it may probably be attributed to inheritance; but not certainly so, as some very distinct American monkeys are thus characterised. The same remark is applicable to the tailless condition of man; for the tail is absent in all the anthropomorphous apes. Nevertheless this character cannot with certainty be attributed to inheritance, as the tail, though not absent, is rudimentary in several other Old World and in some New World species, and is quite absent in several species belonging to the allied group of Lemurs.

Although, as we have now seen, man has no just right to form a separate Order for his own reception, he may 195perhaps claim a distinct Sub-order or Family. Prof. Huxley, in his last work,264 divides the Primates into three Sub-orders; namely, the Anthropidæ with man alone, the Simiadæ including monkeys of all kinds, and the Lemuridæ with the diversified genera of lemurs. As far as differences in certain important points of structure are concerned, man may no doubt rightly claim the rank of a Sub-order; and this rank is too low, if we look chiefly to his mental faculties. Nevertheless, under a genealogical point of view it appears that this rank is too high, and that man ought to form merely a Family, or possibly even only a Sub-family. If we imagine three lines of descent proceeding from a common source, it is quite conceivable that two of them might after the lapse of ages be so slightly changed as still to remain as species of the same genus; whilst the third line might become so greatly modified as to deserve to rank as a distinct Sub-family, Family, or even Order. But in this case it is almost certain that the third line would still retain through inheritance numerous small points of resemblance with the other two lines. Here then would occur the difficulty, at present insoluble, how much weight we ought to assign in our classifications to strongly-marked differences in some few points,—that is to the amount of modification undergone; and how much to close resemblance in numerous unimportant points, as indicating the lines of descent or genealogy. The former alternative is the most obvious, and perhaps the safest, though the latter appears the most correct as giving a truly natural classification.

To form a judgment on this head, with reference to man we must glance at the classification of the 196Simiadæ. This family is divided by almost all naturalists into the Catarhine group, or Old World monkeys, all of which are characterised (as their name expresses) by the peculiar structure of their nostrils and by having four premolars in each jaw; and into the Platyrhine group or New World monkeys (including two very distinct sub-groups), all of which are characterised by differently-constructed nostrils and by having six premolars in each jaw. Some other small differences might be mentioned. Now man unquestionably belongs in his dentition, in the structure of his nostrils, and some other respects, to the Catarhine or Old World division; nor does he resemble the Platyrhines more closely than the Catarhines in any characters, excepting in a few of not much importance and apparently of an adaptive nature. Therefore it would be against all probability to suppose that some ancient New World species had varied, and had thus produced a man-like creature with all the distinctive characters proper to the Old World division; losing at the same time all its own distinctive characters. There can consequently hardly be a doubt that man is an offshoot from the Old World Simian stem; and that under a genealogical point of view, he must be classed with the Catarhine division.265

The anthropomorphous apes, namely the gorilla, chimpanzee, orang, and hylobates, are separated as a distinct sub-group from the other Old World monkeys by most naturalists. I am aware that Gratiolet, relying on the structure of the brain, does not admit the exist197ence of this sub-group, and no doubt it is a broken one; thus the orang, as Mr. St. G. Mivart remarks,266 "is one of the most peculiar and aberrant forms to be found in the Order." The remaining, non-anthropomorphous, Old World monkeys, are again divided by some naturalists into two or three smaller sub-groups; the genus Semnopithecus, with its peculiar sacculated stomach, being the type of one such sub-group. But it appears from M. Gaudry's wonderful discoveries in Attica, that during the Miocene period a form existed there, which connected Semnopithecus and Macacus; and this probably illustrates the manner in which the other and higher groups were once blended together.

If the anthropomorphous apes be admitted to form a natural sub-group, then as man agrees with them, not only in all those characters which he possesses in common with the whole Catarhine group, but in other peculiar characters, such as the absence of a tail and of callosities and in general appearance, we may infer that some ancient member of the anthropomorphous sub-group gave birth to man. It is not probable that a member of one of the other lower sub-groups should, through the law of analogous variation, have given rise to a man-like creature, resembling the higher anthropomorphous apes in so many respects. No doubt man, in comparison with most of his allies, has undergone an extraordinary amount of modification, chiefly in consequence of his greatly developed brain and erect position; nevertheless we should bear in mind that he "is but one of several exceptional forms of Primates."267

Every naturalist, who believes in the principle of 198evolution, will grant that the two main divisions of the Simiadæ, namely the Catarhine and Platyrhine monkeys, with their sub-groups, have all proceeded from some one extremely ancient progenitor. The early descendants of this progenitor, before they had diverged to any considerable extent from each other, would still have formed a single natural group; but some of the species or incipient genera would have already begun to indicate by their diverging characters the future distinctive marks of the Catarhine and Platyrhine divisions. Hence the members of this supposed ancient group would not have been so uniform in their dentition or in the structure of their nostrils, as are the existing Catarhine monkeys in one way and the Platyrhines in another way, but would have resembled in this respect the allied Lemuridæ which differ greatly from each other in the form of their muzzles,268 and to an extraordinary degree in their dentition.

The Catarhine and Platyrhine monkeys agree in a multitude of characters, as is shewn by their unquestionably belonging to one and the same Order. The many characters which they possess in common can hardly have been independently acquired by so many distinct species; so that these characters must have been inherited. But an ancient form which possessed many characters common to the Catarhine and Platyrhine monkeys, and others in an intermediate condition, and some few perhaps distinct from those now present in either group, would undoubtedly have been ranked, if seen by a naturalist, as an ape or monkey.

And as man under a genealogical point of view belongs to the Catarhine or Old World stock, we must conclude, how199ever much the conclusion may revolt our pride, that our early progenitors would have been properly thus designated.269 But we must not fall into the error of supposing that the early progenitor of the whole Simian stock, including man, was identical with, or even closely resembled, any existing ape or monkey.

On the Birthplace and Antiquity of Man.—We are naturally led to enquire where was the birthplace of man at that stage of descent when our progenitors diverged from the Catarhine stock. The fact that they belonged to this stock clearly shews that they inhabited the Old World; but not Australia nor any oceanic island, as we may infer from the laws of geographical distribution. In each great region of the world the living mammals are closely related to the extinct species of the same region. It is therefore probable that Africa was formerly inhabited by extinct apes closely allied to the gorilla and chimpanzee; and as these two species are now man's nearest allies, it is somewhat more probable that our early progenitors lived on the African continent than elsewhere. But it is useless to speculate on this subject, for an ape nearly as large as a man, namely the Dryopithecus of Lartet, which was closely allied to the anthropomorphous Hylobates, existed in Europe during the Upper Miocene period; and since so remote a period the earth has certainly undergone many great revolutions, and there has been ample time for migration on the largest scale.

200

At the period and place, whenever and wherever it may have been, when man first lost his hairy covering, he probably inhabited a hot country; and this would have been favourable for a frugiferous diet, on which, judging from analogy, he subsisted. We are far from knowing how long ago it was when man first diverged from the Catarhine stock; but this may have occurred at an epoch as remote as the Eocene period; for the higher apes had diverged from the lower apes as early as the Upper Miocene period, as shewn by the existence of the Dryopithecus. We are also quite ignorant at how rapid a rate organisms, whether high or low in the scale, may under favourable circumstances be modified: we know, however, that some have retained the same form during an enormous lapse of time. From what we see going on under domestication, we learn that within the same period some of the co-descendants of the same species may be not at all changed, some a little, and some greatly changed. Thus it may have been with man, who has undergone a great amount of modification in certain characters in comparison with the higher apes.

The great break in the organic chain between man and his nearest allies, which cannot be bridged over by any extinct or living species, has often been advanced as a grave objection to the belief that man is descended from some lower form; but this objection will not appear of much weight to those who, convinced by general reasons, believe in the general principle of evolution. Breaks incessantly occur in all parts of the series, some being wide, sharp and defined, others less so in various degrees; as between the orang and its nearest allies—between the Tarsius and the other Lemuridæ—between the elephant and in a more striking manner between the Ornithorhynchus or201 Echidna, and other mammals. But all these breaks depend merely on the number of related forms which have become extinct. At some future period, not very distant as measured by centuries, the civilised races of man will almost certainly exterminate and replace throughout the world the savage races. At the same time the anthropomorphous apes, as Professor Schaaffhausen has remarked,270 will no doubt be exterminated. The break will then be rendered wider, for it will intervene between man in a more civilised state, as we may hope, than the Caucasian, and some ape as low as a baboon, instead of as at present between the negro or Australian and the gorilla.

With respect to the absence of fossil remains, serving to connect man with his ape-like progenitors, no one will lay much stress on this fact, who will read Sir C. Lyell's discussion,271in which he shews that in all the vertebrate classes the discovery of fossil remains has been an extremely slow and fortuitous process. Nor should it be forgotten that those regions which are the most likely to afford remains connecting man with some extinct ape-like creature, have not as yet been searched by geologists.

Lower Stages in the Genealogy of Man.—We have seen that man appears to have diverged from the Catarhine or Old World division of the Simiadæ, after these had diverged from the New World division. We will now endeavour to follow the more remote traces of his genealogy, trusting in the first place to the mutual affinities between the various classes and orders, with some slight aid from the periods, as far as ascertained, 202of their successive appearance on the earth. The Lemuridæ stand below and close to the Simiadæ, constituting a very distinct family of the Primates, or, according to Häckel, a distinct Order. This group is diversified and broken to an extraordinary degree, and includes many aberrant forms. It has, therefore, probably suffered much extinction. Most of the remnants survive on islands, namely in Madagascar and in the islands of the Malayan archipelago, where they have not been exposed to such severe competition as they would have been on well-stocked continents. This group likewise presents many gradations, leading, as Huxley remarks,272 "insensibly from the crown and summit of the animal creation down to creatures from which there is but a step, as it seems, to the lowest, smallest, and least intelligent of the placental mammalia." From these various considerations it is probable that the Simiadæ were originally developed from the progenitors of the existing Lemuridæ; and these in their turn from forms standing very low in the mammalian series.

The Marsupials stand in many important characters below the placental mammals. They appeared at an earlier geological period, and their range was formerly much more extensive than what it now is. Hence the Placentata are generally supposed to have been derived from the Implacentata or Marsupials; not, however, from forms closely like the existing Marsupials, but from their early progenitors. The Monotremata are plainly allied to the Marsupials; forming a third and still lower division in the great mammalian series. They are represented at the present day solely by the Ornithorhynchus and Echidna; and these two forms may 203be safely considered as relics of a much larger group which have been preserved in Australia through some favourable concurrence of circumstances. The Monotremata are eminently interesting, as in several important points of structure they lead towards the class of reptiles.

In attempting to trace the genealogy of the Mammalia, and therefore of man, lower down in the series, we become involved in greater and greater obscurity. He who wishes to see what ingenuity and knowledge can effect, may consult Prof. Häckel's works.273 I will content myself with a few general remarks. Every evolutionist will admit that the five great vertebrate classes, namely, mammals, birds, reptiles, amphibians, and fishes, are all descended from some one prototype; for they have much in common, especially during their embryonic state. As the class of fishes is the most lowly organised and appeared before the others, we may conclude that all the members of the vertebrate kingdom are derived from some fish-like animal, less highly organised than any as yet found in the lowest known formations. The belief that animals so distinct as a monkey or elephant and a humming-bird, a snake, frog, and fish, &c., could all have sprung from the same parents, will appear monstrous to those who have not attended to the recent progress of natural history. For this belief implies the former existence of links closely binding together all these forms, now so utterly unlike.

Nevertheless it is certain that groups of animals have existed, or do now exist, which serve to connect more or less closely the several great vertebrate classes. We have seen that the Ornithorhynchus graduates towards reptiles; and Prof. Huxley has made the remarkable discovery, confirmed by Mr. Cope and others, that the old Dinosaurians are intermediate in many important respects between certain reptiles and certain birds—the latter consisting of the ostrich-tribe (itself evidently a widely-diffused remnant of a larger group) and of the Archeopteryx, that strange Secondary bird having a long tail like that of the lizard. Again, according to Prof. Owen,274 the Ichthyosaurians—great sea-lizards furnished with paddles—present many affinities with fishes, or rather, according to Huxley, with amphibians. This latter class (including in its highest division frogs and toads) is plainly allied to the Ganoid fishes. These latter fishes swarmed during the earlier geological periods, and were constructed on what is called a highly generalised type, that is they presented diversified affinities with other groups of organisms. The amphibians and fishes are also so closely united by the Lepidosiren, that naturalists long disputed in which of these two classes it ought to be placed. The Lepidosiren and some few Ganoid fishes have been preserved from utter extinction by inhabiting our rivers, which are harbours of refuge, bearing the same relation to the great waters of the ocean that islands bear to continents.

Lastly, one single member of the immense and diversified class of fishes, namely the lancelet or amphioxus, is so different from all other fishes, that Häckel maintains that it ought to form a distinct class in the vertebrate kingdom. This fish is remarkable for its 205negative characters; it can hardly be said to possess a brain, vertebral column, or heart, &c.; so that it was classed by the older naturalists amongst the worms. Many years ago Prof. Goodsir perceived that the lancelet presented some affinities with the Ascidians, which are invertebrate, hermaphrodite, marine creatures permanently attached to a support. They hardly appear like animals, and consist of a simple, tough, leathery sack, with two small projecting orifices. They belong to the Molluscoida of Huxley—a lower division of the great kingdom of the Mollusca; but they have recently been placed by some naturalists amongst the Vermes or worms. Their larvæ somewhat resemble tadpoles in shape,275 and have the power of swimming freely about. Some observations lately made by M. Kowalevsky,276 since confirmed by Prof. Kuppfer, will form a discovery of extraordinary interest, if still further extended, as I hear from M. Kowalevsky in Naples he has now effected. The discovery is that the larvæ of Ascidians are related to the Vertebrata, in their manner of development, in the relative position of the nervous system, and in possessing a structure closely like the chorda dorsalis of vertebrate animals. It thus appears, if we may rely on embryology, which has always proved the safest guide in classification, that we have at last gained a clue to the source whence the Vertebrata have 206been derived. We should thus be justified in believing that at an extremely remote period a group of animals existed, resembling in many respects the larvæ of our present Ascidians, which diverged into two great branches—the one retrograding in development and producing the present class of Ascidians, the other rising to the crown and summit of the animal kingdom by giving birth to the Vertebrata.

We have thus far endeavoured rudely to trace the genealogy of the Vertebrata by the aid of their mutual affinities. We will now look to man as he exists; and we shall, I think, be able partially to restore during successive periods, but not in due order of time, the structure of our early progenitors. This can be effected by means of the rudiments which man still retains, by the characters which occasionally make their appearance in him through reversion, and by the aid of the principles of morphology and embryology. The various facts, to which I shall here allude, have been given in the previous chapters. The early progenitors of man were no doubt once covered with hair, both sexes having beards; their ears were pointed and capable of movement; and their bodies were provided with a tail, having the proper muscles. Their limbs and bodies were also acted on by many muscles which now only occasionally reappear, but are normally present in the Quadrumana. The great artery and nerve of the humerus ran through a supra-condyloid foramen. At this or some earlier period, the intestine gave forth a much larger diverticulum or cæcum than that now existing. The foot, judging from the condition of the great toe in the fœtus, was then prehensile; and our progenitors, no doubt, were arboreal in their habits, frequenting some warm, forest-clad land. The males207 were provided with great canine teeth, which served them as formidable weapons.

At a much earlier period the uterus was double; the excreta were voided through a cloaca; and the eye was protected by a third eyelid or nictitating membrane. At a still earlier period the progenitors of man must have been aquatic in their habits; for morphology plainly tells us that our lungs consist of a modified swim-bladder, which once served as a float. The clefts on the neck in the embryo of man show where the branchiæ once existed. At about this period the true kidneys were replaced by the corpora Wolffiana. The heart existed as a simple pulsating vessel; and the chorda dorsalis took the place of a vertebral column. These early predecessors of man, thus seen in the dim recesses of time, must have been as lowly organised as the lancelet or amphioxus, or even still more lowly organised.

There is one other point deserving a fuller notice. It has long been known that in the vertebrate kingdom one sex bears rudiments of various accessory parts, appertaining to the reproductive system, which properly belong to the opposite sex; and it has now been ascertained that at a very early embryonic period both sexes possess true male and female glands. Hence some extremely remote progenitor of the whole vertebrate kingdom appears to have been hermaphrodite or androgynous.277 But here we encounter a singular208difficulty. In the mammalian class the males possess in their vesiculæ prostraticæ rudiments of a uterus with the adjacent passage; they bear also rudiments of mammæ, and some male marsupials have rudiments of a marsupial sack.278 Other analogous facts could be added. Are we, then, to suppose that some extremely ancient mammal possessed organs proper to both sexes, that is, continued androgynous after it had acquired the chief distinctions of its proper class, and therefore after it had diverged from the lower classes of the vertebrate kingdom? This seems improbable in the highest degree; for had this been the case, we might have expected that some few members of the two lower classes, namely fishes279 and amphibians, would still have remained androgynous. We must, on the contrary, believe that when the five vertebrate classes diverged from their common progenitor the sexes had already become separated. To account, however, for male mammals possessing rudiments of the accessory female organs, and for female mammals possessing rudiments of the masculine organs, we need not suppose that their early progenitors were still androgynous after they had assumed their chief mammalian characters. It is quite possible that as the one sex gradually acquired the accessory organs proper to it, some of the successive steps or modifications were transmitted to the opposite sex. When we treat of sexual selection, we shall meet with innumerable instances of this form of transmission,—as in the case of the spurs, plumes, 209and brilliant colours, acquired by male birds for battle or ornament, and transferred to the females in an imperfect or rudimentary condition.

83

The possession by male mammals of functionally imperfect mammary organs is, in some respects, especially curious. The Monotremata have the proper milk-secreting glands with orifices, but no nipples; and as these animals stand at the very base of the mammalian series, it is probable that the progenitors of the class possessed, in like manner, the milk-secreting glands, but no nipples. This conclusion is supported by what is known of their manner of development; for Professor Turner informs me, on the authority of Kölliker and Lauger, that in the embryo the mammary glands can be distinctly traced before the nipples are in the least visible; and it should be borne in mind that the development of successive parts in the individual generally seems to represent and accord with the development of successive beings in the same line of descent. The Marsupials differ from the Monotremata by possessing nipples; so that these organs were probably first acquired by the Marsupials after they had diverged from, and risen above, the Monotremata, and were then transmitted to the placental mammals. No one will suppose that after the Marsupials had approximately acquired their present structure, and therefore at a rather late period in the development of the mammalian series, any of its members still remained androgynous. We seem, therefore, compelled to recur to the foregoing view, and to conclude that the nipples were first developed in the females of some very early marsupial form, and were then, in accordance with a common law of inheritance, transferred in a functionally imperfect condition to the males.

Nevertheless a suspicion has sometimes crossed my210 mind that long after the progenitors of the whole mammalian class had ceased to be androgynous, both sexes might have yielded milk and thus nourished their young; and in the case of the Marsupials, that both sexes might have carried their young in marsupial sacks. This will not appear utterly incredible, if we reflect that the males of syngnathous fishes receive the eggs of the females in their abdominal pouches, hatch them, and afterwards, as some believe, nourish the young;280— that certain other male fishes hatch the eggs within their mouths or branchial cavities;—that certain male toads take the chaplets of eggs from the females and wind them round their own thighs, keeping them there until the tadpoles are born;—that certain male birds undertake the whole duty of incubation, and that male pigeons, as well as the females, feed their nestlings with a secretion from their crops. But the above suspicion first occurred to me from the mammary glands in male mammals being developed so much more perfectly than the rudiments of those other accessory reproductive parts, which are found in the one sex though proper to the other. The mammary glands and nipples, as they exist in male mammals, can indeed hardly be called rudimentary; they are simply not fully developed and not functionally active. They are sympathetically affected under the influence of certain diseases, like the same organs in the female. At birth they often secrete a few drops of milk; and they have 211been known occasionally in man and other mammals to become well developed, and to yield a fair supply of milk. Now if we suppose that during a former prolonged period male mammals aided the females in nursing their offspring, and that afterwards from some cause, as from a smaller number of young being produced, the males ceased giving this aid, disuse of the organs during maturity would lead to their becoming inactive; and from two well-known principles of inheritance this state of inactivity would probably be transmitted to the males at the corresponding age of maturity. But at all earlier ages these organs would be left unaffected, so that they would be equally well developed in the young of both sexes.

Conclusion.—The best definition of advancement or progress in the organic scale ever given, is that by Von Baer; and this rests on the amount of differentiation and specialisation of the several parts of the same being, when arrived, as I should be inclined to add, at maturity. Now as organisms have become slowly adapted by means of natural selection for diversified lines of life, their parts will have become, from the advantage gained by the division of physiological labour, more and more differentiated and specialised for various functions. The same part appears often to have been modified first for one purpose, and then long afterwards for some other and quite distinct purpose; and thus all the parts are rendered more and more complex. But each organism will still retain the general type of structure of the progenitor from which it was aboriginally derived. In accordance with this view it seems, if we turn to geological evidence, that organisation on the whole has advanced throughout the world by slow and interrupted steps. In the great212 kingdom of the Vertebrata it has culminated in man. It must not, however, be supposed that groups of organic beings are always supplanted and disappear as soon as they have given birth to other and more perfect groups. The latter, though victorious over their predecessors, may not have become better adapted for all places in the economy of nature. Some old forms appear to have survived from inhabiting protected sites, where they have not been exposed to very severe competition; and these often aid us in constructing our genealogies, by giving us a fair idea of former and lost populations. But we must not fall into the error of looking at the existing members of any lowly-organised group as perfect representatives of their ancient predecessors.

The most ancient progenitors in the kingdom of the Vertebrata, at which we are able to obtain an obscure glance, apparently consisted of a group of marine animals,281 resembling the larvæ of existing Ascidians. These animals probably gave rise to a group of fishes, as lowly organised as the lancelet; and from these the Ganoids, and other fishes like the Lepidosiren, must have been developed. From such fish a very small advance would 213carry us on to the amphibians. We have seen that birds and reptiles were once intimately connected together; and the Monotremata now, in a slight degree, connect mammals with reptiles. But no one can at present say by what line of descent the three higher and related classes, namely, mammals, birds, and reptiles, were derived from either of the two lower vertebrate classes, namely amphibians and fishes. In the class of mammals the steps are not difficult to conceive which led from the ancient Monotremata to the ancient Marsupials; and from these to the early progenitors of the placental mammals. We may thus ascend to the Lemuridæ; and the interval is not wide from these to the Simiadæ. The Simiadæ then branched off into two great stems, the New World and Old World monkeys; and from the latter, at a remote period, Man, the wonder and glory of the Universe, proceeded.

Thus we have given to man a pedigree of prodigious length, but not, it may be said, of noble quality. The world, it has often been remarked, appears as if it had long been preparing for the advent of man; and this, in one sense is strictly true, for he owes his birth to a long line of progenitors. If any single link in this chain had never existed, man would not have been exactly what he now is. Unless we wilfully close our eyes, we may, with our present knowledge, approximately recognise our parentage; nor need we feel ashamed of it. The most humble organism is something much higher than the inorganic dust under our feet; and no one with an unbiassed mind can study any living creature, however humble, without being struck with enthusiasm at its marvellous structure and properties.

CHAPTER VII.
On the Races of Man.

The nature and value of specific characters—Application to the races of man—Arguments in favour of, and opposed to, ranking the so-called races of man as distinct species—Sub-species—Monogenists and polygenists—Convergence of character—Numerous points of resemblance in body and mind between the most distinct races of man—The state of man when he first spread over the earth—Each race not descended from a single pair—The extinction of races—The formation of races—The effects of crossing—Slight influence of the direct action of the conditions of life—Slight or no influence of natural selection—Sexual selection.

It is not my intention here to describe the several so-called races of men; but to inquire what is the value of the differences between them under a classificatory point of view, and how they have originated. In determining whether two or more allied forms ought to be ranked as species or varieties, naturalists are practically guided by the following considerations; namely, the amount of difference between them, and whether such differences relate to few or many points of structure, and whether they are of physiological importance; but more especially whether they are constant. Constancy of character is what is chiefly valued and sought for by naturalists. Whenever it can be shewn, or rendered probable, that the forms in question have remained distinct for a long period, this becomes an argument of much weight in favour of treating them as species. Even a slight degree of sterility between any two forms when first crossed, or in their offspring, is generally considered as a decisive215 test of their specific distinctness; and their continued persistence without blending within the same area, is usually accepted as sufficient evidence, either of some degree of mutual sterility, or in the case of animals of some repugnance to mutual pairing.

Independently of blending from intercrossing, the complete absence, in a well-investigated region, of varieties linking together any two closely-allied forms, is probably the most important of all the criterions of their specific distinctness; and this is a somewhat different consideration from mere constancy of character, for two forms may be highly variable and yet not yield intermediate varieties. Geographical distribution is often unconsciously and sometimes consciously brought into play; so that forms living in two widely separated areas, in which most of the other inhabitants are specifically distinct, are themselves usually looked at as distinct; but in truth this affords no aid in distinguishing geographical races from so-called good or true species.

Now let us apply these generally-admitted principles to the races of man, viewing him in the same spirit as a naturalist would any other animal. In regard to the amount of difference between the races, we must make some allowance for our nice powers of discrimination gained by the long habit of observing ourselves. In India, as Elphinstone remarks,282although a newly-arrived European cannot at first distinguish the various native races, yet they soon appear to him extremely dissimilar; and the Hindoo cannot at first perceive any difference between the several European nations. Even the most distinct races of man, with the exception of certain negro tribes, are much more like each other in form 216than would at first be supposed. This is well shewn by the French photographs in the Collection Anthropologique du Muséum of the men belonging to various races, the greater number of which, as many persons to whom I have shown them have remarked, might pass for Europeans. Nevertheless, these men if seen alive would undoubtedly appear very distinct, so that we are clearly much influenced in our judgment by the mere colour of the skin and hair, by slight differences in the features, and by expression.

There is, however, no doubt that the various races, when carefully compared and measured, differ much from each other,—as in the texture of the hair, the relative proportions of all parts of the body,[283] the capacity of the lungs, the form and capacity of the skull, and even in the convolutions of the brain.[284] But it would be an endless task to specify the numerous points of structural difference. The races differ also in constitution, in acclimatisation, and in liability to certain diseases. Their mental characteristics are likewise very distinct; chiefly as it would appear in their emotional, but partly in their intellectual, faculties. Every one who has had the opportunity of comparison, must have been struck with the contrast between the taciturn, even morose, aborigines of S. America and the light-hearted, talkative negroes. There is a nearly similar contrast between the Malays and the Papuans,[285] who live 217under the same physical conditions, and are separated from each other only by a narrow space of sea.

We will first consider the arguments which may be advanced in favour of classing the races of man as distinct species, and then those on the other side. If a naturalist, who had never before seen such beings, were to compare a Negro, Hottentot, Australian, or Mongolian, he would at once perceive that they differed in a multitude of characters, some of slight and some of considerable importance. On inquiry he would find that they were adapted to live under widely different climates, and that they differed somewhat in bodily constitution and mental disposition. If he were then told that hundreds of similar specimens could be brought from the same countries, he would assuredly declare that they were as good species as many to which he had been in the habit of affixing specific names. This conclusion would be greatly strengthened as soon as he had ascertained that these forms had all retained the same character for many centuries; and that negroes, apparently identical with existing negroes, had lived at least 4000 years ago.[286] He would also hear from an excellent observer, 218Dr. Lund,[287] that the human skulls found in the caves of Brazil, entombed with many extinct mammals, belonged to the same type as that now prevailing throughout the American Continent.

Our naturalist would then perhaps turn to geographical distribution, and he would probably declare that forms differing not only in appearance, but fitted for the hottest and dampest or driest countries, as well as for the arctic regions, must be distinct species. He might appeal to the fact that no one species in the group next to man, namely the Quadrumana, can resist a low temperature or any considerable change of climate; and that those species which come nearest to man have never been reared to maturity, even under the temperate climate of Europe. He would be deeply impressed with the fact, first noticed by Agassiz,[288] that the different races of man are distributed over the world in the same zoological provinces, as those inhabited by undoubtedly distinct species and genera of mammals. This is manifestly the case with the Australian, Mongolian, and Negro races of man; in a less well-marked manner with the Hottentots; but plainly with the Papuans and Malays, who are separated, as Mr. Wallace has shewn, by nearly the same line which divides the great Malayan and Australian zoological provinces. The aborigines of America range throughout the Continent; and this at first appears opposed to the above rule, for most of the productions of the Southern and Northern halves differ widely; yet some few living forms, as the 219opossum, range from the one into the other, as did formerly some of the gigantic Edentata. The Esquimaux, like other Arctic animals, extend round the whole polar regions. It should be observed that the mammalian forms which inhabit the several zoological provinces, do not differ from each other in the same degree; so that it can hardly be considered as an anomaly that the Negro differs more, and the American much less, from the other races of man than do the mammals of the same continents from those of the other provinces. Man, it may be added, does not appear to have aboriginally inhabited any oceanic island; and in this respect he resembles the other members of his class.

In determining whether the varieties of the same kind of domestic animal should be ranked as specifically distinct, that is, whether any of them are descended from distinct wild species, every naturalist would lay much stress on the fact, if established, of their external parasites being specifically distinct. All the more stress would be laid on this fact, as it would be an exceptional one, for I am informed by Mr. Denny that the most different kinds of dogs, fowls, and pigeons, in England, are infested by the same species of Pediculi or lice. Now Mr. A. Murray has carefully examined the Pediculi collected in different countries from the different races of man;[289] and he finds that they differ, not only in colour, but in the structure of their claws and limbs. In every case in which numerous specimens were obtained the differences were constant. The surgeon of a whaling ship in the Pacific assured me that when the Pediculi, with which some Sandwich Islanders on board swarmed, strayed on to the bodies of the English sailors, they died in the course of three or four days. These Pediculi 220were darker coloured and appeared different from those proper to the natives of Chiloe in South America, of which he gave me specimens. These, again, appeared larger and much softer than European lice. Mr. Murray procured four kinds from Africa, namely from the Negroes of the Eastern and Western coasts, from the Hottentots and Caffres; two kinds from the natives of Australia; two from North, and two from South America. In these latter cases it may be presumed that the Pediculi came from natives inhabiting different districts. With insects slight structural differences, if constant, are generally esteemed of specific value: and the fact of the races of man being infested by parasites, which appear to be specifically distinct, might fairly be urged as an argument that the races themselves ought to be classed as distinct species.

Our supposed naturalist having proceeded thus far in his investigation, would next inquire whether the races of men, when crossed, were in any degree sterile. He might consult the work[290] of a cautious and philosophical observer, Professor Broca; and in this he would find good, evidence that some races were quite fertile together; but evidence of an opposite nature in regard to other races. Thus it has been asserted that the native women of Australia and Tasmania rarely produce children to European men; the evidence, however, on this head has now been shewn to be almost valueless. The half-castes are killed by the pure blacks; and an account has lately been published of eleven half-caste youths murdered and burnt at the same time, whose remains were found by the police.[291] Again, it has often 221been said that when mulattoes intermarry they produce few children; on the other hand, Dr. Bachman of Charlestown[292] positively asserts that he has known mulatto families which have intermarried for several generations, and have continued on an average as fertile as either pure whites or pure blacks. Inquiries formerly made by Sir C. Lyell on this subject led him, as he informs me, to the same conclusion. In the United States the census for the year 1854 included, according to Dr. Bachman, 405,751 mulattoes; and this number, considering all the circumstances of the case, seems small; but it may partly be accounted for by the degraded and anomalous position of the class, and by the profligacy of the women. A certain amount of absorption of mulattoes into negroes must always be in progress; and this would lead to an apparent diminution of the former. The inferior vitality of mulattoes is spoken of in a trustworthy work[293] as a well-known phenomenon; but this is a different consideration from their lessened fertility; and can hardly be advanced as a proof of the specific distinctness of the parent races. No doubt both animal and vegetable hybrids, when produced from extremely distinct species, are liable to premature death; but the parents of mulattoes cannot be put under the category of extremely distinct species. The common Mule, so notorious for long life and vigour, and yet so sterile, shews how little necessary connection 222there is in hybrids between lessened fertility and vitality: other analogous cases could be added.

Even if it should hereafter be proved that all the races of men were perfectly fertile together, he who was inclined from other reasons to rank them as distinct species, might with justice argue that fertility and sterility are not safe criterions of specific distinctness. We know that these qualities are easily affected by changed conditions of life or by close inter-breeding, and that they are governed by highly complex laws, for instance that of the unequal fertility of reciprocal crosses between the same two species. With forms which must be ranked as undoubted species, a perfect series exists from those which are absolutely sterile when crossed, to those which are almost or quite fertile. The degrees of sterility do not coincide strictly with the degrees of difference in external structure or habits of life. Man in many respects may be compared with those animals which have long been domesticated, and a large body of evidence can be advanced in favour of the Pallasian doctrine294 that domestication tends to eliminate the 223sterility which is so general a result of the crossing of species in a state of nature. From these several considerations, it may be justly urged that the perfect fertility of the intercrossed races of man, if established, would not absolutely preclude us from ranking them as distinct species.

Independently of fertility, the character of the offspring from a cross has sometimes been thought to afford evidence whether the parent-forms ought to be ranked as species or varieties; but after carefully studying the evidence, I have come to the conclusion that no general rules of this kind can be trusted. Thus with mankind the offspring of distinct races resemble in all respects the offspring of true species and of varieties. This is shewn, for instance, by the manner in which the characters of both parents are blended, and by one form absorbing another through repeated crosses. In this latter case the progeny both of crossed species and varieties retain for a long period a tendency to revert to their ancestors, especially to that one which is prepotent in transmission. When any character has suddenly appeared in a race or species as the result of a 224single act of variation, as is general with monstrosities,295 and this race is crossed with another not thus characterised, the characters in question do not commonly appear in a blended condition in the young, but are transmitted to them either perfectly developed or not at all. As with the crossed races of man cases of this kind rarely or never occur, this may be used as an argument against the view suggested by some ethnologists, namely that certain characters, for instance the blackness of the negro, first appeared as a sudden variation or sport. Had this occurred, it is probable that mulattoes would often have been born, either completely black or completely white.

We have now seen that a naturalist might feel himself fully justified in ranking the races of man as distinct species; for he has found that they are distinguished by many differences in structure and constitution, some being of importance. These differences have, also, remained nearly constant for very long periods of time. He will have been in some degree influenced by the enormous range of man, which is a great anomaly in the class of mammals, if mankind be viewed as a single species. He will have been struck with the distribution of the several so-called races, in accordance with that of other undoubtedly distinct species of mammals. Finally he might urge that the mutual fertility of all the races has not as yet been fully proved; and even if proved would not be an absolute proof of their specific identity.

On the other side of the question, if our supposed naturalist were to enquire whether the forms of man kept distinct like ordinary species, when mingled to225gether in large numbers in the same country, he would immediately discover that this was by no means the case. In Brazil he would behold an immense mongrel population of Negroes and Portuguese; in Chiloe and other parts of South America, he would behold the whole population consisting of Indians and Spaniards blended in various degrees.296 In many parts of the same continent he would meet with the most complex crosses between Negroes, Indians, and Europeans; and such triple crosses afford the severest test, judging from the vegetable kingdom, of the mutual fertility of the parent-forms. In one island of the Pacific he would find a small population of mingled Polynesian and English blood; and in the Viti Archipelago a population of Polynesians and Negritos crossed in all degrees. Many analogous cases could be added, for instance, in South Africa. Hence the races of man are not sufficiently distinct to co-exist without fusion; and this it is, which in all ordinary cases affords the usual test of specific distinctness.

Our naturalist would likewise be much disturbed as soon as he perceived that the distinctive characters of every race of man were highly variable. This strikes every one when he first beholds the negro-slaves in Brazil, who have been imported from all parts of Africa. The same remark holds good with the Polynesians, and with many other races. It may be doubted whether any character can be named which is distinctive of a race and is constant. Savages, even within the limits of the same tribe, are not nearly so uniform in character, as has often been said. Hottentot women offer certain 226peculiarities, more strongly marked than those occurring in any other race, but these are known not to be of constant occurrence. In the several American tribes, colour and hairyness differ considerably; as does colour to a certain degree, and the shape of the features greatly, in the Negroes of Africa. The shape of the skull varies much in some races;297 and so it is with every other character. Now all naturalists have learnt by dearly-bought experience, how rash it is to attempt to define species by the aid of inconstant characters.

But the most weighty of all the arguments against treating the races of man as distinct species, is that they graduate into each other, independently in many cases, as far as we can judge, of their having intercrossed. Man has been studied more carefully than any other organic being, and yet there is the greatest possible diversity amongst capable judges whether he should be classed as a single species or race, or as two (Virey), as three (Jacquinot), as four (Kant), five (Blumenbach), six (Buffon), seven (Hunter), eight (Agassiz), eleven (Pickering), fifteen (Bory St. Vincent), sixteen (Desmoulins), twenty-two (Morton), sixty (Crawfurd), or as sixty-three, according to Burke.298 This diversity of judgment does not prove that the races ought not to be ranked as species, but it shews that they graduate into each other, and that it is hardly possible to discover clear distinctive characters between them.

Every naturalist who has had the misfortune to under227take the description of a group of highly varying organisms, has encountered cases (I speak after experience) precisely like that of man; and if of a cautious disposition, he will end by uniting all the forms which graduate into each other as a single species; for he will say to himself that he has no right to give names to objects which he cannot define. Cases of this kind occur in the Order which includes man, namely in certain genera of monkeys; whilst in other genera, as in Cercopithecus, most of the species can be determined with certainty. In the American genus

Cebus, the various forms are ranked by some naturalists as species, by others as mere geographical races. Now if numerous specimens of Cebus were collected from all parts of South America, and those forms which at present appear to be specifically distinct, were found to graduate into each other by close steps, they would be ranked by most naturalists as mere varieties or races; and thus the greater number of naturalists have acted with respect to the races of man. Nevertheless it must be confessed that there are forms, at least in the vegetable kingdom,[299] which we cannot avoid naming as species, but which are connected together, independently of intercrossing, by numberless gradations.

Some naturalists have lately employed the term "sub-species" to designate forms which possess many of the characteristics of true species, but which hardly deserve so high a rank. Now if we reflect on the weighty arguments, above given, for raising the races of man to the dignity of species, and the insuperable difficulties on the other side in defining them, the term "sub-species"[228] might here be used with much propriety. But from long habit the term "race" will perhaps always be employed. The choice of terms is only so far important as it is highly desirable to use, as far as that may be possible, the same terms for the same degrees of difference. Unfortunately this is rarely possible; for within the same family the larger genera generally include closely-allied forms, which can be distinguished only with much difficulty, whilst the smaller genera include forms that are perfectly distinct; yet all must equally be ranked as species. So again the species within the same large genus by no means resemble each other to the same degree: on the contrary, in most cases some of them can be arranged in little groups round other species, like satellites round planets.[300]

The question whether mankind consists of one or several species has of late years been much agitated by anthropologists, who are divided into two schools of monogenists and polygenists. Those who do not admit the principle of evolution, must look at species either as separate creations or as in some manner distinct entities; and they must decide what forms to rank as species by the analogy of other organic beings which are commonly thus received. But it is a hopeless endeavour to decide this point on sound grounds, until some definition of the term "species" is generally accepted; and the definition must not include an element which cannot possibly be ascertained, such as an act of creation. We might as well attempt without any definition to decide whether a certain number of houses should be called a village, or town, or city. We have a practical illustration of the difficulty in the never-[229]ending doubts whether many closely-allied mammals, birds, insects, and plants, which represent each other in North America and Europe, should be ranked species or geographical races; and so it is with the productions of many islands situated at some little distance from the nearest continent.

Those naturalists, on the other hand, who admit the principle of evolution, and this is now admitted by the greater number of rising men, will feel no doubt that all the races of man are descended from a single primitive stock; whether or not they think fit to designate them as distinct species, for the sake of expressing their amount of difference.[301] With our domestic animals the question whether the various races have arisen from one or more species is different. Although all such races, as well as all the natural species within the same genus, have undoubtedly sprung from the same primitive stock, yet it is a fit subject for discussion, whether, for instance, all the domestic races of the dog have acquired their present differences since some one species was first domesticated and bred by man; or whether they owe some of their characters to inheritance from distinct species, which had already been modified in a state of nature. With mankind no such question can arise, for he cannot be said to have been domesticated at any particular period.

When the races of man diverged at an extremely remote epoch from their common progenitor, they will have differed but little from each other, and been few in number; consequently they will then, as far as their distinguishing characters are concerned, have had less claim to rank as distinct species, than the existing so230-called races. Nevertheless such early races would perhaps have been ranked by some naturalists as distinct species, so arbitrary is the term, if their differences, although extremely slight, had been more constant than at present, and had not graduated into each other.

It is, however, possible, though far from probable, that the early progenitors of man might at first have diverged much in character, until they became more unlike each other than are any existing races; but that subsequently, as suggested by Vogt,302 they converged in character. When man selects for the same object the offspring of two distinct species, he sometimes induces, as far as general appearance is concerned, a considerable amount of convergence. This is the case, as shewn by Von Nathusius,303 with the improved breeds of pigs, which are descended from two distinct species; and in a less well-marked manner with the improved breeds of cattle. A great anatomist, Gratiolet, maintains that the anthropomorphous apes do not form a natural sub-group; but that the orang is a highly developed gibbon or semnopithecus; the chimpanzee a highly developed macacus; and the gorilla a highly developed mandrill. If this conclusion, which rests almost exclusively on brain-characters, be admitted, we should have a case of convergence at least in external characters, for the anthropomorphous apes are certainly more like each other in many points than they are to other apes. All analogical resemblances, as of a whale to a fish, may indeed be said to be cases of convergence; but this term has never been applied to superficial and adaptive resemblances. It 231would be extremely rash in most cases to attribute to convergence close similarity in many points of structure in beings which had once been widely different. The form of a crystal is determined solely by the molecular forces, and it is not surprising that dissimilar substances should sometimes assume the same form; but with organic beings we should bear in mind that the form of each depends on an infinitude of complex relations, namely on the variations which have arisen, these being due to causes far too intricate to be followed out,—on the nature of the variations which have been preserved, and this depends on the surrounding physical conditions, and in a still higher degree on the surrounding organisms with which each has come into competition,—and lastly, on inheritance (in itself a fluctuating element) from innumerable progenitors, all of which have had their forms determined through equally complex relations. It appears utterly incredible that two organisms, if differing in a marked manner, should ever afterwards converge so closely as to lead to a near approach to identity throughout their whole organisation. In the case of the convergent pigs above referred to, evidence of their descent from two primitive stocks is still plainly retained, according to Von Nathusius, in certain bones of their skulls. If the races of man were descended, as supposed by some naturalists, from two or more distinct species, which had differed as much, or nearly as much, from each other, as the orang differs from the gorilla, it can hardly be doubted that marked differences in the structure of certain bones would still have been discoverable in man as he now exists.

Although the existing races of man differ in many respects, as in colour, hair, shape of skull, proportions of the body, &c., yet if their whole organisation be taken232 into consideration they are found to resemble each other closely in a multitude of points. Many of these points are of so unimportant or of so singular a nature, that it is extremely improbable that they should have been independently acquired by aboriginally distinct species or races. The same remark holds good with equal or greater force with respect to the numerous points of mental similarity between the most distinct races of man. The American aborigines, Negroes and Europeans differ as much from each other in mind as any three races that can be named; yet I was incessantly struck, whilst living with the Fuegians on board the "Beagle," with the many little traits of character, shewing how similar their minds were to ours; and so it was with a full-blooded negro with whom I happened once to be intimate.

He who will carefully read Mr. Tylor's and Sir J. Lubbock's interesting works304 can hardly fail to be deeply impressed with the close similarity between the men of all races in tastes, dispositions and habits. This is shewn by the pleasure which they all take in dancing, rude music, acting, painting, tattooing, and otherwise decorating themselves,—in their mutual comprehension of gesture-language—and, as I shall be able to shew in a future essay, by the same expression in their features, and by the same inarticulate cries, when they are excited by various emotions. This similarity, or rather identity, is striking, when contrasted with the different expressions which may be observed in distinct species of monkeys. There is good evidence that the art of shooting with bows and arrows has not been handed down from any common progenitor of 233mankind, yet the stone arrow-heads, brought from the most distant parts of the world and manufactured at the most remote periods, are, as Nilsson has shewn,305 almost identical; and this fact can only be accounted for by the various races having similar inventive or mental powers. The same observation has been made by archæologists306 with respect to certain widely-prevalent ornaments, such as zigzags, &c.; and with respect to various simple beliefs and customs, such as the burying of the dead under megalithic structures. I remember observing in South America,307 that there, as in so many other parts of the world, man has generally chosen the summits of lofty hills, on which to throw up piles of stones, either for the sake of recording some remarkable event, or for burying his dead.

Now when naturalists observe a close agreement in numerous small details of habits, tastes and dispositions between two or more domestic races, or between nearly-allied natural forms, they use this fact as an argument that all are descended from a common progenitor who was thus endowed; and consequently that all should be classed under the same species. The same argument may be applied with much force to the races of man.

As it is improbable that the numerous and unimportant points of resemblance between the several races of man in bodily structure and mental faculties (I do not here refer to similar customs) should all have been independently acquired, they must have been inherited from progenitors who were thus characterised. We thus gain some insight into the early state of man, 234before he had spread step by step over the face of the earth. The spreading of man to regions widely separated by the sea, no doubt, preceded any considerable amount of divergence of character in the several races; for otherwise we should sometimes meet with the same race in distinct continents; and this is never the case. Sir J. Lubbock, after comparing the arts now practised by savages in all parts of the world, specifies those which man could not have known, when he first wandered from his original birthplace; for if once learnt they would never have been forgotten.308 He thus shews that "the spear, which is but a development of the knife-point, and the club, which is but a long hammer, are the only things left." He admits, however, that the art of making fire probably had already been discovered, for it is common to all the races now existing, and was known to the ancient cave-inhabitants of Europe. Perhaps the art of making rude canoes or rafts was likewise known; but as man existed at a remote epoch, when the land in many places stood at a very different level, he would have been able, without the aid of canoes, to have spread widely.

Sir J. Lubbock further remarks how improbable it is that our earliest ancestors could have "counted as high as ten, considering that so many races now in existence cannot get beyond four." Nevertheless, at this early period, the intellectual and social faculties of man could hardly have been inferior in any extreme degree to those now possessed by the lowest savages; otherwise primeval man could not have been so eminently successful in the struggle for life, as proved by his early and wide diffusion.

From the fundamental differences between certain 235languages, some philologists have inferred that when man first became widely diffused he was not a speaking animal; but it may be suspected that languages, far less perfect than any now spoken, aided by gestures, might have been used, and yet have left no traces on subsequent and more highly-developed tongues. Without the use of some language, however imperfect, it appears doubtful whether man's intellect could have risen to the standard implied by his dominant position at an early period.

Whether primeval man, when he possessed very few arts of the rudest kind, and when his power of language was extremely imperfect, would have deserved to be called man, must depend on the definition which we employ. In a series of forms graduating insensibly from some ape-like creature to man as he now exists, it would be impossible to fix on any definite point when the term "man" ought to be used. But this is a matter of very little importance. So again it is almost a matter of indifference whether the so-called races of man are thus designated, or are ranked as species or sub-species; but the latter term appears the most appropriate. Finally, we may conclude that when the principles of evolution are generally accepted, as they surely will be before long, the dispute between the monogenists and the polygenists will die a silent and unobserved death.

One other question ought not to be passed over without notice, namely, whether, as is sometimes assumed, each sub-species or race of man has sprung from a single pair of progenitors. With our domestic animals a new race can readily be formed from a single pair possessing some new character, or even from a single individual thus characterised, by carefully match236ing the varying offspring; but most of our races have been formed, not intentionally from a selected pair, but unconsciously by the preservation of many individuals which have varied, however slightly, in some useful or desired manner. If in one country stronger and heavier horses, and in another country lighter and fleeter horses, were habitually preferred, we may feel sure that two distinct sub-breeds would, in the course of time, be produced, without any particular pairs or individuals having been separated and bred from in either country. Many races have been thus formed, and their manner of formation is closely analogous with that of natural species. We know, also, that the horses which have been brought to the Falkland Islands have become, during successive generations, smaller and weaker, whilst those which have run wild on the Pampas have acquired larger and coarser heads; and such changes are manifestly due, not to any one pair, but to all the individuals having been subjected to the same conditions, aided, perhaps, by the principle of reversion. The new sub-breeds in none of these cases are descended from any single pair, but from many individuals which have varied in different degrees, but in the same general manner; and we may conclude that the races of man have been similarly produced, the modifications being either the direct result of exposure to different conditions, or the indirect result of some form of selection. But to this latter subject we shall presently return.

On the Extinction of the Races of Man.—The partial and complete extinction of many races and sub-races of man are historically known events. Humboldt saw in South America a parrot which was the sole living creature that could speak the language of a lost tribe.237 Ancient monuments and stone implements found in all parts of the world, of which no tradition is preserved by the present inhabitants, indicate much extinction. Some small and broken tribes, remnants of former races, still survive in isolated and generally mountainous districts. In Europe the ancient races were all, according to Schaaffhausen,309 "lower in the scale than the rudest living savages;" they must therefore have differed, to a certain extent, from any existing race. The remains described by Professor Broca310 from Les Eyzies, though they unfortunately appear to have belonged to a single family, indicate a race with a most singular combination of low or simious and high characteristics, and is "entirely different from any other race, ancient or modern, that we have ever heard of." It differed, therefore, from the quaternary race of the caverns of Belgium.

Unfavourable physical conditions appear to have had but little effect in the extinction of races.311 Man has long lived in the extreme regions of the North, with no wood wherewith to make his canoes or other implements, and with blubber alone for burning and giving him warmth, but more especially for melting the snow. In the Southern extremity of America the Fuegians survive without the protection of clothes, or of any building worthy to be called a hovel. In South Africa the aborigines wander over the most arid plains, where dangerous beasts abound. Man can withstand the deadly influence of the Terai at the foot of the Himalaya, and the pestilential shores of tropical Africa.

238

Extinction follows chiefly from the competition of tribe with tribe, and race with race. Various checks are always in action, as specified in a former chapter, which serve to keep down the numbers of each savage tribe,—such as periodical famines, the wandering of the parents and the consequent deaths of infants, prolonged suckling, the stealing of women, wars, accidents, sickness, licentiousness, especially infanticide, and, perhaps, lessened fertility from less nutritious food, and many hardships. If from any cause any one of these checks is lessened, even in a slight degree, the tribe thus favoured will tend to increase; and when one of two adjoining tribes becomes more numerous and powerful than the other, the contest is soon settled by war, slaughter, cannibalism, slavery, and absorption. Even when a weaker tribe is not thus abruptly swept away, if it once begins to decrease, it generally goes on decreasing until it is extinct.312

When civilised nations come into contact with barbarians the struggle is short, except where a deadly climate gives its aid to the native race. Of the causes which lead to the victory of civilised nations, some are plain and some very obscure. We can see that the cultivation of the land will be fatal in many ways to savages, for they cannot, or will not, change their habits. New diseases and vices are highly destructive; and it appears that in every nation a new disease causes much death, until those who are most susceptible to its destructive influence are gradually weeded out;313 and so it may be with the evil effects from spirituous liquors, as well as with the unconquerably strong taste for them shewn by so many savages. It further appears, mysterious as is 239the fact, that the first meeting of distinct and separated people generates disease.314 Mr. Sproat, who in Vancouver Island closely attended to the subject of extinction, believes that changed habits of life, which always follow from the advent of Europeans, induces much ill-health. He lays, also, great stress on so trifling a cause as that the natives become "bewildered and dull by the new life around them; they lose the motives for exertion, and get no new ones in their place."315

The grade of civilisation seems a most important element in the success of nations which come in competition. A few centuries ago Europe feared the inroads of Eastern barbarians; now, any such fear would be ridiculous. It is a more curious fact, that savages did not formerly waste away, as Mr. Bagehot has remarked, before the classical nations, as they now do before modern civilised nations; had they done so, the old moralists would have mused over the event; but there is no lament in any writer of that period over the perishing barbarians.316

Although the gradual decrease and final extinction of the races of man is an obscure problem, we can see that it depends on many causes, differing in different places and at different times. It is the same difficult problem as that presented by the extinction of one of the higher animals—of the fossil horse, for instance, which disappeared from South America, soon afterwards to be replaced, within the same districts, by countless troops of the Spanish horse. The New Zealander seems 240conscious of this parallelism, for he compares his future fate with that of the native rat almost exterminated by the European rat. The difficulty, though great to our imagination, and really great if we wish to ascertain the precise causes, ought not to be so to our reason, as long as we keep steadily in mind that the increase of each species and each race is constantly hindered by various checks; so that if any new check, or cause of destruction, even a slight one, be superadded, the race will surely decrease in number; and as it has everywhere been observed that savages are much opposed to any change of habits, by which means injurious checks could be counterbalanced, decreasing numbers will sooner or later lead to extinction; the end, in most cases, being promptly determined by the inroads of increasing and conquering tribes.

On the Formation of the Races of Man.—It may be premised that when we find the same race, though broken up into distinct tribes, ranging over a great area, as over America, we may attribute their general resemblance to descent from a common stock. In some cases the crossing of races already, distinct has led to the formation of new races. The singular fact that Europeans and Hindoos, who belong to the same Aryan stock and speak a language fundamentally the same, differ widely in appearance, whilst Europeans differ but little from Jews, who belong to the Semitic stock and speak quite another language, has been accounted for by Broca317 through the Aryan branches having been largely crossed during their wide diffusion by various indigenous tribes. When two races in close contact 241cross, the first result is a heterogeneous mixture: thus Mr. Hunter, in describing the Santali or hill-tribes of India, says that hundreds of imperceptible gradations may be traced "from the black, squat tribes of the mountains to the tall olive-coloured Brahman, with his intellectual brow, calm eyes, and high but narrow head;" so that it is necessary in courts of justice to ask the witnesses whether they are Santalis or Hindoos.318 Whether a heterogeneous people, such as the inhabitants of some of the Polynesian islands, formed by the crossing of two distinct races, with few or no pure members left, would ever become homogeneous, is not known from direct evidence. But as with our domesticated animals, a crossed breed can certainly, in the course of a few generations, be fixed and made uniform by careful selection,319 we may infer that the free and prolonged intercrossing during many generations of a heterogeneous mixture would supply the place of selection, and overcome any tendency to reversion, so that a crossed race would ultimately become homogeneous, though it might not partake in an equal degree of the characters of the two parent-races.

Of all the differences between the races of man, the colour of the skin is the most conspicuous and one of the best marked. Differences of this kind, it was formerly thought, could be accounted for by long exposure under different climates; but Pallas first shewed that this view is not tenable, and he has been followed by almost all anthropologists.[320] The view has been 242rejected chiefly because the distribution of the variously coloured races, most of whom must have long inhabited their present homes, does not coincide with corresponding differences of climate. Weight must also be given to such cases as that of the Dutch families, who, as we hear on excellent authority,[321] have not undergone the least change of colour, after residing for three centuries in South Africa. The uniform appearance in various parts of the world of gypsies and Jews, though the uniformity of the latter has been somewhat exaggerated,[322] is likewise an argument on the same side. A very damp or a very dry atmosphere has been supposed to be more influential in modifying the colour of the skin than mere heat; but as D'Orbigny in South America, and Livingstone in Africa, arrived at diametrically opposite conclusions with respect to dampness and dryness, any conclusion on this head must be considered as very doubtful.[323]

Various facts, which I have elsewhere given, prove that the colour of the skin and hair is sometimes correlated in a surprising manner with a complete immunity from the action of certain vegetable poisons and from the attacks of certain parasites. Hence it occurred to me, that negroes and other dark races might have acquired their dark tints by the darker individuals escaping during a long series of generations from the deadly influence of the miasmas of their native countries.

I afterwards found that the same idea had long ago 243occurred to Dr. Wells.[324] That negroes, and even mulattoes, are almost completely exempt from the yellow-fever, which is so destructive in tropical America, has long been known.[325] They likewise escape to a large extent the fatal intermittent fevers that prevail along, at least, 2600 miles of the shores of Africa, and which annually cause one-fifth of the white settlers to die, and another fifth to return home invalided.[326] This immunity in the negro seems to be partly inherent, depending on some unknown peculiarity of constitution, and partly the result of acclimatisation. Pouchet[327] states that the negro regiments, borrowed from the Viceroy of Egypt for the Mexican war, which had been recruited near the Soudan, escaped the yellow-fever almost equally well with the negroes originally brought from various parts of Africa, and accustomed to the climate of the West Indies. That acclimatisation plays a part is shewn by the many cases in which negroes, after having resided for some time in a colder climate, have become to a certain extent liable to tropical fevers.[328] The nature of the climate under which the white races have long resided, likewise has some influence on them; for during the fearful epidemic of yellow-fever in Demerara during 1837, Dr. Blair found that the death-rate of the immigrants was proportional 244to the latitude of the country whence they had come. With the negro the immunity, as far as it is the result of acclimatisation, implies exposure during a prodigious length of time; for the aborigines of tropical America, who have resided there from time immemorial, are not exempt from yellow-fever; and the Rev. B. Tristram states, that there are districts in Northern Africa which the native inhabitants are compelled annually to leave, though the negroes can remain with safety.

That the immunity of the negro is in any degree correlated with the colour of his skin is a mere conjecture: it may be correlated with some difference in his blood, nervous system, or other tissues. Nevertheless, from the facts above alluded to, and from some connection apparently existing between complexion and a tendency to consumption, the conjecture seemed to me not improbable. Consequently I endeavoured, with but little success,[329] to ascertain how far it held good. The 245late Dr. Daniell, who had long lived on the West Coast of Africa, told me that he did not believe in any such relation. He was himself unusually fair, and had withstood the climate in a wonderful manner. When he first arrived as a boy on the coast, an old and experienced negro chief predicted from his appearance that this would prove the case. Dr. Nicholson, of Antigua, after having attended to this subject, wrote to me that he did not think that dark-coloured Europeans escaped the yellow-fever better than those that were light-coloured. Mr. J. M. Harris altogether denies330 that Europeans with dark hair withstand a hot climate better than other men; on the contrary, experience has taught him in making a selection of men for service on the coast of Africa, to choose those with red hair. As far, therefore, as these slight indications serve, there seems no foundation for the hypothesis, which has been accepted by several writers, that the colour of the black races may have resulted from darker and darker individuals having survived in greater numbers, during their exposure to the fever-generating miasmas of their native countries.

Although with our present knowledge we cannot account for the strongly-marked differences in colour between the races of man, either through correlation with constitutional peculiarities, or through the direct action of climate; yet we must not quite ignore the 246latter agency, for there is good reason to believe that some inherited effect is thus produced.331

We have seen in our third chapter that the conditions of life, such as abundant food and general comfort, affect in a direct manner the development of the bodily frame, the effects being transmitted. Through the combined influences of climate and changed habits of life, European settlers, in the United States undergo, as is generally admitted, a slight but extraordinarily rapid change of appearance. There is, also, a considerable body of evidence shewing that in the Southern States the house-slaves of the third generation present a markedly different appearance from the field-slaves.332

If, however, we look to the races of man, as distributed over the world, we must infer that their characteristic differences cannot be accounted for by the direct action of different conditions of life, even after exposure to them for an enormous period of time. The Esquimaux live exclusively on animal food; they are clothed in thick fur, and are exposed to intense cold and to prolonged darkness; yet they do not differ in any extreme degree from the inhabitants of Southern China, who live entirely on vegetable food and are exposed almost naked to a hot, glaring climate. The unclothed Fuegians live on the marine productions of their inhospitable shores; the Botocudos of Brazil wander 247about the hot forests of the interior and live chiefly on vegetable productions; yet these tribes resemble each other so closely that the Fuegians on board the "Beagle" were mistaken by some Brazilians for Botocudos. The Botocudos again, as well as the other inhabitants of tropical America, are wholly different from the Negroes who inhabit the opposite shores of the Atlantic, are exposed to a nearly similar climate, and follow nearly the same habits of life.

Nor can the differences between the races of man be accounted for, except to a quite insignificant degree, by the inherited effects of the increased or decreased use of parts. Men who habitually live in canoes, may have their legs somewhat stunted; those who inhabit lofty regions have their chests enlarged; and those who constantly use certain sense-organs have the cavities in which they are lodged somewhat increased in size, and their features consequently a little modified. With civilised nations, the reduced size of the jaws from lessened use, the habitual play of different muscles serving to express different emotions, and the increased size of the brain from greater intellectual activity, have together produced a considerable effect on their general appearance in comparison with savages.333 It is also possible that increased bodily stature, with no corresponding increase in the size of the brain, may have given to some races (judging from the previously adduced cases of the rabbits) an elongated skull of the dolichocephalic type.

Lastly, the little-understood principle of correlation will almost certainly have come into action, as in the case of great muscular development and strongly pro248jecting supra-orbital ridges. It is not improbable that the texture of the hair, which differs much in the different races, may stand in some kind of correlation with the structure of the skin; for the colour of the hair and skin are certainly correlated, as is its colour and texture with the Mandans.334 The colour of the skin and the odour emitted by it are likewise in some manner connected. With the breeds of sheep the number of hairs within a given space and the number of the excretory pores stand in some relation to each other.335 If we may judge from the analogy of our domesticated animals, many modifications of structure in man probably come under this principle of correlated growth.

We have now seen that the characteristic differences between the races of man cannot be accounted for in a satisfactory manner by the direct action of the conditions of life, nor by the effects of the continued use of parts, nor through the principle of correlation. We are therefore led to inquire whether slight individual differences, to which man is eminently liable, may not have been preserved and augmented during a long series of generations through natural selection. But here we are at once met by the objection that beneficial variations alone can be thus preserved; and as far as we are enabled to judge (although always liable to error on this head) not one of the external differences between the races of man are of any direct or 249special service to him. The intellectual and moral or social faculties must of course be excepted from this remark; but differences in these faculties can have had little or no influence on external characters. The variability of all the characteristic differences between the races, before referred to, likewise indicates that these differences cannot be of much importance; for, had they been important, they would long ago have been either fixed and preserved, or eliminated. In this respect man resembles those forms, called by naturalists protean or polymorphic, which have remained extremely variable, owing, as it seems, to their variations being of an indifferent nature, and consequently to their having escaped the action of natural selection.

We have thus far been baffled in all our attempts to account for the differences between the races of man; but there remains one important agency, namely Sexual Selection, which appears to have acted as powerfully on man, as on many other animals. I do not intend to assert that sexual selection will account for all the differences between the races. An unexplained residuum is left, about which we can in our ignorance only say, that as individuals are continually born with, for instance, heads a little rounder or narrower, and with noses a little longer or shorter, such slight differences might become fixed and uniform, if the unknown agencies which induced them were to act in a more constant manner, aided by long-continued intercrossing. Such modifications come under the provisional class, alluded to in our fourth chapter, which for the want of a better term have been called spontaneous variations. Nor do I pretend that the effects of sexual selection can be indicated with scientific precision; but it can be shewn that it would be an inexplicable fact if man had not been modified by this agency, which has250 acted so powerfully on innumerable animals, both high and low in the scale. It can further be shewn that the differences between the races of man, as in colour, hairyness, form of features, &c., are of the nature which it might have been expected would have been acted on by sexual selection. But in order to treat this subject in a fitting manner, I have found it necessary to pass the whole animal kingdom in review; I have therefore devoted to it the Second Part of this work. At the close

I shall return to man, and, after attempting to shew how far he has been modified through sexual selection, will give a brief summary of the chapters in this First Part.

Part II.
SEXUAL SELECTION.

CHAPTER VIII.
Principles of Sexual Selection.

Secondary sexual characters—Sexual selection—Manner of action—Excess of males—Polygamy—The male alone generally modified through sexual selection—Eagerness of the male—Variability of the male—Choice exerted by the female—Sexual compared with natural selection—Inheritance, at corresponding periods of life, at corresponding seasons of the year, and as limited by sex—Relations between the several forms of inheritance—Causes why one sex and the young are not modified through sexual selection—Supplement on the proportional numbers of the two sexes throughout the animal kingdom—On the limitation of the numbers of the two sexes through natural selection.

With animals which have their sexes separated, the males necessarily differ from the females in their organs of reproduction; and these afford the primary sexual characters. But the sexes often differ in what Hunter has called secondary sexual characters, which are not directly connected with the act of reproduction; for instance, in the male possessing certain organs of sense or locomotion, of which the female is quite destitute, or in having them more highly-developed, in order that he may readily find or reach her; or again, in the male having special organs of prehension so as to hold her securely. These latter organs of infinitely diversified kinds graduate into, and in some cases can hardly be distinguished from, those which are commonly ranked as primary, such as the complex appendages at the apex of the abdomen in male insects. Unless indeed254 we confine the term "primary" to the reproductive glands, it is scarcely possible to decide, as far as the organs of prehension are concerned, which ought to be called primary and which secondary.

The female often differs from the male in having organs for the nourishment or protection of her young, as the mammary glands of mammals, and the abdominal sacks of the marsupials. The male, also, in some few cases differs from the female in possessing analogous organs, as the receptacles for the ova possessed by the males of certain fishes, and those temporarily developed in certain male frogs. Female bees have a special apparatus for collecting and carrying pollen, and their ovipositor is modified into a sting for the defence of their larvæ and the community. In the females of many insects the ovipositor is modified in the most complex manner for the safe placing of the eggs. Numerous similar cases could be given, but they do not here concern us. There are, however, other sexual differences quite disconnected with the primary organs with which we are more especially concerned—such as the greater size, strength, and pugnacity of the male, his weapons of offence or means of defence against rivals, his gaudy colouring and various ornaments, his power of song, and other such characters.

Besides the foregoing primary and secondary sexual differences, the male and female sometimes differ in structures connected with different habits of life, and not at all, or only indirectly, related to the reproductive functions. Thus the females of certain flies (Culicidæ and Tabanidæ) are blood-suckers, whilst the males live on flowers and have their mouths destitute of mandibles.336 The males alone of certain moths and of some 255crustaceans (e.g. Tanais) have imperfect, closed mouths, and cannot feed. The Complemental males of certain cirripedes live like epiphytic plants either on the female or hermaphrodite form, and are destitute of a mouth and prehensile limbs. In these cases it is the male which has been modified and has lost certain important organs, which the other members of the same group possess. In other cases it is the female which has lost such parts; for instance, the female glow-worm is destitute of wings, as are many female moths, some of which never leave their cocoons. Many female parasitic crustaceans have lost their natatory legs. In some weevil-beetles (Curculionidæ) there is a great difference between the male and female in the length of the rostrum or snout;337 but the meaning of this and of many analogous differences, is not at all understood. Differences of structure between the two sexes in relation to different habits of life are generally confined to the lower animals; but with some few birds the beak of the male differs from that of the female. No doubt in most, but apparently not in all these cases, the differences are indirectly connected with the propagation of the species: thus a female which has to nourish a multitude of ova will require more food than the male, and consequently will require special means for procuring it. A male animal which lived for a very short time might without detriment lose through disuse its organs for procuring food; but he would retain his locomotive organs in a perfect state, so that he might reach the female. The female, on the other hand, might safely lose her organs for flying, swimming, 256or walking, if she gradually acquired habits which rendered such powers useless.

We are, however, here concerned only with that kind of selection, which I have called sexual selection. This depends on the advantage which certain individuals have over other individuals of the same sex and species, in exclusive relation to reproduction. When the two sexes differ in structure in relation to different habits of life, as in the cases above mentioned, they have no doubt been modified through natural selection, accompanied by inheritance limited to one and the same sex. So again the primary sexual organs, and those for nourishing or protecting the young, come under this same head; for those individuals which generated or nourished their offspring best, would leave, cæteris paribus, the greatest number to inherit their superiority; whilst those which generated or nourished their offspring badly, would leave but few to inherit their weaker powers. As the male has to search for the female, he requires for this purpose organs of sense and locomotion, but if these organs are necessary for the other purposes of life, as is generally the case, they will have been developed through natural selection. When the male has found the female he sometimes absolutely requires prehensile organs to hold her; thus Dr. Wallace informs me that the males of certain moths cannot unite with the females if their tarsi or feet are broken. The males of many oceanic crustaceans have their legs and antennæ modified in an extraordinary manner for the prehension of the female; hence we may suspect that owing to these animals being washed about by the waves of the open sea, they absolutely require these organs in order to propagate their kind, and if so their development will have been the result of ordinary or natural selection.

When the two sexes follow exactly the same habits257 of life, and the male has more highly developed sense or locomotive organs than the female, it may be that these in their perfected state are indispensable to the male for finding the female; but in the vast majority of cases, they serve only to give one male an advantage over another, for the less well-endowed males, if time were allowed them, would succeed in pairing with the females; and they would in all other respects, judging from the structure of the female, be equally well adapted for their ordinary habits of life. In such cases sexual selection must have come into action, for the males have acquired their present structure, not from being better fitted to survive in the struggle for existence, but from having gained an advantage over other males, and from having transmitted this advantage to their male offspring alone. It was the importance of this distinction which led me to designate this form of selection as sexual selection. So again, if the chief service rendered to the male by his prehensile organs is to prevent the escape of the female before the arrival of other males, or when assaulted by them, these organs will have been perfected through sexual selection, that is by the advantage acquired by certain males over their rivals. But in most cases it is scarcely possible to distinguish between the effects of natural and sexual selection. Whole chapters could easily be filled with details on the differences between the sexes in their sensory, locomotive, and prehensile organs. As, however, these structures are not more interesting than others adapted for the ordinary purposes of life, I shall almost pass them over, giving only a few instances under each class.

There are many other structures and instincts which must have been developed through sexual selection—such as the weapons of offence and the means of defence[258] possessed by the males for fighting with and driving away their rivals—their courage and pugnacity—their ornaments of many kinds—their organs for producing vocal or instrumental music—and their glands for emitting odours; most of these latter structures serving only to allure or excite the female. That these characters are the result of sexual and not of ordinary selection is clear, as unarmed, unornamented, or unattractive males would succeed equally well in the battle for life and in leaving a numerous progeny, if better endowed males were not present. We may infer that this would be the case, for the females, which are unarmed and unornamented, are able to survive and procreate their kind. Secondary sexual characters of the kind just referred to, will be fully discussed in the following chapters, as they are in many respects interesting, but more especially as they depend on the will, choice, and rivalry of the individuals of either sex. When we behold two males fighting for the possession of the female, or several male birds displaying their gorgeous plumage, and performing the strangest antics before an assembled body of females, we cannot doubt that, though led by instinct, they know what they are about, and consciously exert their mental and bodily powers.

In the same manner as man can improve the breed of his game-cocks by the selection of those birds which are victorious in the cockpit, so it appears that the strongest and most vigorous males, or those provided with the best weapons, have prevailed under nature, and have led to the improvement of the natural breed or species. Through repeated deadly contests, a slight degree of variability, if it led to some advantage, however slight, would suffice for the work of sexual selection; and it is certain that secondary sexual characters[259] are eminently variable. In the same manner as man can give beauty, according to his standard of taste, to his male poultry—can give to the Sebright bantam a new and elegant plumage, an erect and peculiar carriage—so it appears that in a state of nature female birds, by having long selected the more attractive males, have added to their beauty. No doubt this implies powers of discrimination and taste on the part of the female which will at first appear extremely improbable; but I hope hereafter to shew that this is not the case.

From our ignorance on several points, the precise manner in which sexual selection acts is to a certain extent uncertain. Nevertheless if those naturalists who already believe in the mutability of species, will read the following chapters, they will, I think, agree with me that sexual selection has played an important part in the history of the organic world. It is certain that with almost all animals there is a struggle between the males for the possession of the female. This fact is so notorious that it would be superfluous to give instances. Hence the females, supposing that their mental capacity sufficed for the exertion of a choice, could select one out of several males. But in numerous cases it appears as if it had been specially arranged that there should be a struggle between many males. Thus with migratory birds, the males generally arrive before the females at their place of breeding, so that many males are ready to contend for each female. The bird-catchers assert that this is invariably the case with the nightingale and blackcap, as I am informed by Mr. Jenner Weir, who confirms the statement with respect to the latter species.

Mr. Swaysland of Brighton, who has been in the habit, during the last forty years, of catching our migratory birds on their first arrival, writes to me that he has260 never known the females of any species to arrive before their males. During one spring he shot thirty-nine males of Ray's wagtail (Budytes Raii) before he saw a single female. Mr. Gould has ascertained by dissection, as he informs me, that male snipes arrive in this country before the females. In the case of fish, at the period when the salmon ascend our rivers, the males in large numbers are ready to breed before the females. So it apparently is with frogs and toads. Throughout the great class of insects the males almost always emerge from the pupal state before the other sex, so that they generally swarm for a time before any females can be seen.338 The cause of this difference between the males and females in their periods of arrival and maturity is sufficiently obvious. Those males which annually first migrated into any country, or which in the spring were first ready to breed, or were the most eager, would leave the largest number of offspring; and these would tend to inherit similar instincts and constitutions. On the whole there can be no doubt that with almost all animals, in which the sexes are separate, there is a constantly recurrent struggle between the males for the possession of the females.

Our difficulty in regard to sexual selection lies in understanding how it is that the males which conquer other males, or those which prove the most attractive to the females, leave a greater number of offspring to inherit their superiority than the beaten and less 261attractive males. Unless this result followed, the characters which gave to certain males an advantage over others, could not be perfected and augmented through sexual selection. When the sexes exist in exactly equal numbers, the worst-endowed males will ultimately find females (excepting where polygamy prevails), and leave as many offspring, equally well fitted for their general habits of life, as the best-endowed males. From various facts and considerations, I formerly inferred that with most animals, in which secondary sexual characters were well developed, the males considerably exceeded the females in number; and this does hold good in some few cases. If the males were to the females as two to one, or as three to two, or even in a somewhat lower ratio, the whole affair would be simple; for the better-armed or more attractive males would leave the largest number of offspring. But after investigating, as far as possible, the numerical proportions of the sexes, I do not believe that any great inequality in number commonly exists. In most cases sexual selection appears to have been effective in the following manner.

Let us take any species, a bird for instance, and divide the females inhabiting a district into two equal bodies: the one consisting of the more vigorous and better-nourished individuals, and the other of the less vigorous and healthy. The former, there can be little doubt, would be ready to breed in the spring before the others; and this is the opinion of Mr. Jenner Weir, who has during many years carefully attended to the habits of birds. There can also be no doubt that the most vigorous, healthy, and best-nourished females would on an average succeed in rearing the largest number of offspring. The males, as we have seen, are generally ready to breed before the females; of the males the262 strongest, and with some species the best armed, drive away the weaker males; and the former would then unite with the more vigorous and best-nourished females, as these are the first to breed. Such vigorous pairs would surely rear a larger number of offspring than the retarded females, which would be compelled, supposing the sexes to be numerically equal, to unite with the conquered and less powerful males; and this is all that is wanted to add, in the course of successive generations, to the size, strength and courage of the males, or to improve their weapons.

But in a multitude of cases the males which conquer other males, do not obtain possession of the females, independently of choice on the part of the latter. The courtship of animals is by no means so simple and short an affair as might be thought. The females are most excited by, or prefer pairing with, the more ornamented males, or those which are the best songsters, or play the best antics; but it is obviously probable, as has been actually observed in some cases, that they would at the same time prefer the more vigorous and lively males.339 Thus the more vigorous females, which are the first to breed, will have the choice of many males; and though they may not always select the strongest or best armed, they will select those which are vigorous and well armed, and in other respects the most attractive. Such early pairs would have the same advantage in rearing offspring on the female side as above explained, and nearly the same advantage on the male side. And this apparently has sufficed during a long course of generations to add not only to the strength and fighting-powers of 263the males, but likewise to their various ornaments or other attractions.

In the converse and much rarer case of the males selecting particular females, it is plain that those which were the most vigorous and had conquered others, would have the freest choice; and it is almost certain that they would select vigorous as well as attractive females. Such pairs would have an advantage in rearing offspring, more especially if the male had the power to defend the female during the pairing-season, as occurs with some of the higher animals, or aided in providing for the young. The same principles would apply if both sexes mutually preferred and selected certain individuals of the opposite sex; supposing that they selected not only the more attractive, but likewise the more vigorous individuals.

Numerical Proportion of the Two Sexes.—I have remarked that sexual selection would be a simple affair if the males considerably exceeded in number the females. Hence I was led to investigate, as far as I could, the proportions between the two sexes of as many animals as possible; but the materials are scanty. I will here give only a brief abstract of the results, retaining the details for a supplementary discussion, so as not to interfere with the course of my argument. Domesticated animals alone afford the opportunity of ascertaining the proportional numbers at birth; but no records have been specially kept for this purpose. By indirect means, however, I have collected a considerable body of statistical data, from which it appears that with most of our domestic animals the sexes are nearly equal at birth. Thus with race-horses, 25,560 births have been recorded during twenty-one years, and the male births have been to the female births as 99·7 to 100. With greyhounds the264 inequality is greater than with any other animal, for during twelve years, out of 6878 births, the male births have been as 110·1 to 100 female births. It is, however, in some degree doubtful whether it is safe to infer that the same proportional numbers would hold good under natural conditions as under domestication; for slight and unknown differences in the conditions affect to a certain extent the proportion of the sexes. Thus with mankind, the male births in England are as 104·5, in Russia as 108·9, and with the Jews of Livornia as 120 to 100 females. The proportion is also mysteriously affected by the circumstance of the births being legitimate or illegitimate.

For our present purpose we are concerned with the proportion of the sexes, not at birth, but at maturity, and this adds another element of doubt; for it is a well ascertained fact that with man a considerably larger proportion of males than of females die before or during birth, and during the first few years of infancy. So it almost certainly is with male lambs, and so it may be with the males of other animals. The males of some animals kill each other by fighting; or they drive each other about until they become greatly emaciated. They must, also, whilst wandering about in eager search for the females, be often exposed to various dangers. With many kinds of fish the males are much smaller than the females, and they are believed often to be devoured by the latter or by other fishes. With some birds the females appear to die in larger proportion than the males: they are also liable to be destroyed on their nests, or whilst in charge of their young. With insects the female larvæ are often larger than those of the males, and would consequently be more likely to be devoured: in some cases the mature females are less active and less rapid in their movements than the males,265 and would not be so well able to escape from danger. Hence, with animals in a state of nature, in order to judge of the proportions of the sexes at maturity, we must rely on mere estimation; and this, except perhaps when the inequality is strongly marked, is but little trustworthy. Nevertheless, as far as a judgment can be formed, we may conclude from the facts given in the supplement, that the males of some few mammals, of many birds, of some fish and insects, considerably exceed in number the females.

The proportion between the sexes fluctuates slightly during successive years: thus with race-horses, for every 100 females born, the males varied from 1O7.1 in one year to 92.6 in another year, and with greyhounds from 116.3 to 95.3. But had larger numbers been tabulated throughout a more extensive area than England, these fluctuations would probably have disappeared; and such as they are, they would hardly suffice to lead under a state of nature to the effective action of sexual selection. Nevertheless with some few wild animals, the proportions seem, as shewn in the supplement, to fluctuate either during different seasons or in different localities in a sufficient degree to lead to such action. For it should be observed that any advantage gained during certain years or in certain localities by those males which were able to conquer other males, or were the most attractive to the females, would probably be transmitted to the offspring and would not subsequently be eliminated. During the succeeding seasons, when from the equality of the sexes every male was everywhere able to procure a female, the stronger or more attractive males previously produced would still have at least as good a chance of leaving offspring as the less strong or less attractive.

Polygamy.—The practice of polygamy leads to the266 same results as would follow from an actual inequality in the number of the sexes; for if each male secures two or more females, many males will not be able to pair; and the latter assuredly will be the weaker or less attractive individuals. Many mammals and some few birds are polygamous, but with animals belonging to the lower classes I have found no evidence of this habit. The intellectual powers of such animals are, perhaps, not sufficient to lead them to collect and guard a harem of females. That some relation exists between polygamy and the development of secondary sexual characters, appears nearly certain; and this supports the view that a numerical preponderance of males would be eminently favourable to the action of sexual selection. Nevertheless many animals, especially birds, which are strictly monogamous, display strongly-marked secondary sexual characters; whilst some few animals, which are polygamous, are not thus characterised.

We will first briefly run through the class of mammals, and then turn to birds. The gorilla seems to be a polygamist, and the male differs considerably from the female; so it is with some baboons which live in herds containing twice as many adult females as males. In South America the Mycetes caraya presents well-marked sexual differences in colour, beard, and vocal organs, and the male generally lives with two or three wives: the male of the Cebus capucinus differs somewhat from the female, and appears to be polygamous.340 Little is known on this head with respect to most other monkeys, but some species are strictly monogamous. The ruminants are eminently polygamous, and they 267more frequently present sexual differences than almost any other group of mammals, especially in their weapons, but likewise in other characters. Most deer, cattle, and sheep are polygamous; as are most antelopes, though some of the latter are monogamous. Sir Andrew Smith, in speaking of the antelopes of South Africa, says that in herds of about a dozen there was rarely more than one mature male. The Asiatic Antilope saiga appears to be the most inordinate polygamist in the world; for Pallas341 states that the male drives away all rivals, and collects a herd of about a hundred, consisting of females and kids: the female is hornless and has softer hair, but does not otherwise differ much from the male. The horse is polygamous, but, except in his greater size and in the proportions of his body, differs but little from the mare. The wild boar, in his great tusks and some other characters, presents well-marked sexual characters; in Europe and in India he leads a solitary life, except during the breeding-season; but at this season he consorts in India with several females, as Sir W. Elliot, who has had large experience in observing this animal, believes: whether this holds good in Europe is doubtful, but is supported by some statements. The adult male Indian elephant, like the boar, passes much of his time in solitude; but when associating with others, "it is rare to find," as Dr. Campbell states, "more than one male with a whole herd of females." The larger males expel or kill the smaller and weaker ones. The male differs from the female by his immense tusks and greater size, strength, and endurance; so great is the difference in these latter 268respects, that the males when caught are valued at twenty per cent. above the females.342 With other pachydermatous animals the sexes differ very little or not at all, and they are not, as far as known, polygamists. Hardly a single species amongst the Cheiroptera and Edentata, or in the great Orders of the Rodents and Insectivora, presents well-developed secondary sexual differences; and I can find no account of any species being polygamous, excepting, perhaps, the common rat, the males of which, as some rat-catchers affirm, live with several females.

The lion in South Africa, as I hear from Sir Andrew Smith, sometimes lives with a single female, but generally with more than one, and, in one case, was found with as many as five females, so that he is polygamous. He is, as far as I can discover, the sole polygamist in the whole group of the terrestrial Carnivora, and he alone presents well-marked sexual characters. If, however, we turn to the marine Carnivora, the case is widely different; for many species of seals offer, as we shall hereafter see, extraordinary sexual differences, and they are eminently polygamous. Thus the male sea-elephant of the Southern Ocean, always possesses, according to Péron, several females, and the sea-lion of Forster is said to be surrounded by from twenty to thirty females. In the North, the male sea-bear of Steller is accompanied by even a greater number of females.

With respect to birds, many species, the sexes of which differ greatly from each other, are certainly monogamous. In Great Britain we see well-marked sexual differences in, for instance, the wild-duck which pairs with a single female, with the common blackbird, 269and with the bullfinch which is said to pair for life. So it is, as I am informed by Mr. Wallace, with the Chatterers or Cotingidæ of South America, and numerous other birds. In several groups I have not been able to discover whether the species are polygamous or monogamous. Lesson says that birds of paradise, so remarkable for their sexual differences, are polygamous, but Mr. Wallace doubts whether he had sufficient evidence. Mr. Salvin informs me that he has been led to believe that humming-birds are polygamous. The male widow-bird; remarkable for his caudal plumes, certainly seems to be a polygamist.343 I have been assured by Mr. Jenner Weir and by others, that three starlings not rarely frequent the same nest; but whether this is a case of polygamy or polyandry has not been ascertained.

The Gallinaceæ present almost as strongly marked sexual differences as birds of paradise or humming-birds, and many of the species are, as is well known, polygamous; others being strictly monogamous. What a contrast is presented between the sexes by the polygamous peacock or pheasant, and the monogamous guinea-fowl or partridge! Many similar cases could be given, as in the grouse tribe, in which the males of the polygamous capercailzie and black-cock differ greatly from the females; whilst the sexes of the monogamous red grouse and ptarmigan differ very little. Amongst the Cursores, no great number of species offer strongly-marked sexual differences, except the bustards, and the great bustard (Otis tarda), is said to 270be polygamous. With the Grallatores, extremely few species differ sexually, but the ruff (Machetes pugnax) affords a strong exception, and this species is believed by Montagu to be a polygamist. Hence it appears that with birds there often exists a close relation between polygamy and the development of strongly-marked sexual differences. On asking Mr. Bartlett, at the Zoological Gardens, who has had such large experience with birds, whether the male tragopan (one of the Gallinaceæ) was polygamous, I was struck by his answering, "I do not know, but should think so from his splendid colours."

It deserves notice that the instinct of pairing with a single female is easily lost under domestication. The wild-duck is strictly monogamous, the domestic-duck highly polygamous. The Rev. W. D. Fox informs me that with some half-tamed wild-ducks, kept on a large pond in his neighbourhood, so many mallards were shot by the gamekeeper that only one was left for every seven or eight females; yet unusually large broods were reared. The guinea-fowl is strictly monogamous; but Mr. Fox finds that his birds succeed best when he keeps one cock to two or three hens.344 Canary-birds pair in a state of nature, but the breeders in England successfully put one male to four or five females; nevertheless the first female, as Mr. Fox has been assured, is alone treated as the wife, she and her young ones being fed by him; the others are treated as concubines. I have noticed these cases, as it renders it in some degree probable that monogamous species, in a state of nature, might readily become either temporarily or permanently polygamous.

271

With respect to reptiles and fishes, too little is known of their habits to enable us to speak of their marriage arrangements. The stickle-back Gasterosteus, however, is said to be a polygamist;345 and the male during the breeding-season differs conspicuously from the female.

To sum up on the means through which, as far as we can judge, sexual selection has led to the development of secondary sexual characters. It has been shewn that the largest number of vigorous offspring will be reared from the pairing of the strongest and best-armed males, which have conquered other males, with the most vigorous and best-nourished females, which are the first to breed in the spring. Such females, if they select the more attractive, and at the same time vigorous, males, will rear a larger number of offspring than the retarded females, which must pair with the less vigorous and less attractive males. So it will be if the more vigorous males select the more attractive and at the same time healthy and vigorous females; and this will especially hold good if the male defends the female, and aids in providing food for the young. The advantage thus gained by the more vigorous pairs in rearing a larger number of offspring has apparently sufficed to render sexual selection efficient. But a large preponderance in number of the males over the females would be still more efficient; whether the preponderance was only occasional and local, or permanent; whether it occurred at birth, or subsequently from the greater destruction of the females; or whether it indirectly followed from the practice of polygamy.

The Male generally more modified than the Female.—Throughout the animal kingdom, when the sexes differ 272from each other in external appearance, it is the male which, with rare exceptions, has been chiefly modified; for the female still remains more like the young of her own species, and more like the other members of the same group. The cause of this seems to lie in the males of almost all animals having stronger passions than the females. Hence it is the males that fight together and sedulously display their charms before the females; and those which are victorious transmit their superiority to their male offspring. Why the males do not transmit their characters to both sexes will hereafter be considered. That the males of all mammals eagerly pursue the females is notorious to every one. So it is with birds; but many male birds do not so much pursue the female, as display their plumage, perform strange antics, and pour forth their song, in her presence. With the few fish which have been observed, the male seems much more eager than the female; and so it is with alligators, and apparently with Batrachians. Throughout the enormous class of insects, as Kirby remarks,346 "the law is, that the male shall seek the female." With spiders and crustaceans, as I hear from two great authorities, Mr. Blackwall and Mr. C. Spence Bate, the males are more active and more erratic in their habits than the females. With insects and crustaceans, when the organs of sense or locomotion are present in the one sex and absent in the other, or when, as is frequently the case, they are more highly developed in the one than the other, it is almost invariably the male, as far as I can discover, which retains such organs, or has them most developed; and this shews that the male is the more active member in the courtship of the sexes.347

273

The female, on the other hand, with the rarest exception, is less eager than the male. As the illustrious Hunter348 long ago observed, she generally "requires to be courted;" she is coy, and may often be seen endeavouring for a long time to escape from the male. Every one who has attended to the habits of animals will be able to call to mind instances of this kind. Judging from various facts, hereafter to be given, and from the results which may fairly be attributed to sexual selection, the female, though comparatively passive, generally exerts some choice and accepts one male in preference to others. Or she may accept, as appearances would sometimes lead us to believe, not the male which is the most attractive to her, but the one which is the least distasteful. The exertion of some choice on the part of the female seems almost as general a law as the eagerness of the male.

We are naturally led to enquire why the male in so many and such widely distinct classes has been rendered more eager than the female, so that he searches for her and plays the more active part in courtship. It would be no advantage and some loss of power if both sexes were mutually to search for each other; but why should the male almost always be the seeker? With plants, the ovules after fertilisation have to be nourished for a time; hence the pollen is necessarily brought to the female organs—being placed on the stigma, through the agency of insects or of the wind, 274or by the spontaneous movements of the stamens; and with the Algæ, &c., by the locomotive power of the antherozooids. With lowly-organised animals permanently affixed to the same spot and having their sexes separate, the male element is invariably brought to the female; and we can see the reason; for the ova, even if detached before being fertilised and not requiring subsequent nourishment or protection, would be, from their larger relative size, less easily transported than the male element. Hence plants349 and many of the lower animals are, in this respect, analogous. In the case of animals not affixed to the same spot, but enclosed within a shell with no power of protruding any part of their bodies, and in the case of animals having little power of locomotion, the males must trust the fertilising element to the risk of at least a short transit through the waters of the sea. It would, therefore, be a great advantage to such animals, as their organisation became perfected, if the males when ready to emit the fertilising element, were to acquire the habit of approaching the female as closely as possible. The males of various lowly-organised animals having thus aboriginally acquired the habit of approaching and seeking the females, the same habit would naturally be transmitted to their more highly developed male descendants; and in order that they should become efficient seekers, they would have to be endowed with strong passions. The acquirement of such passions would naturally follow from the more eager males leaving a larger number of offspring than the less eager.

The great eagerness of the male has thus indirectly 275led to the much more frequent development of secondary sexual characters in the male than in the female. But the development of such characters will have been much aided, if the conclusion at which I arrived after studying domesticated animals, can be trusted, namely, that the male is more liable to vary than the female. I am aware how difficult it is to verify a conclusion of this kind. Some slight evidence, however, can be gained by comparing the two sexes in mankind, as man has been more carefully observed than any other animal. During the Novara Expedition350 a vast number of measurements of various parts of the body in different races were made, and the men were found in almost every case to present a greater range of variation than the women; but I shall have to recur to this subject in a future chapter. Mr. J. Wood,351 who has carefully attended to the variation of the muscles in man, puts in italics the conclusion that "the greatest number of abnormalities in each subject is found in the males." He had previously remarked that "altogether in 102 subjects the varieties of redundancy were found to be half as many again as in females, contrasting widely with the greater frequency of deficiency in females before described." Professor Macalister like wise remarks352 that variations in the muscles "are probably more common in males than females." Certain muscles which are not normally present in mankind are also more frequently developed in the male than in the female sex, although exceptions to this rule 276are said to occur. Dr. Burt Wilder353 has tabulated the cases of 152 individuals with supernumerary digits, of which 86 were males, and 39, or less than half, females; the remaining 27 being of unknown sex. It should not, however, be overlooked that women would more frequently endeavour to conceal a deformity of this kind than men. Whether the large proportional number of deaths of the male offspring of man and apparently of sheep, compared with the female offspring, before, during, and shortly after birth (see supplement), has any relation to a stronger tendency in the organs of the male to vary and thus to become abnormal in structure or function, I will not pretend to conjecture.

In various classes of animals a few exceptional cases occur, in which the female instead of the male has acquired well pronounced secondary sexual characters, such as brighter colours, greater size, strength, or pugnacity. With birds, as we shall hereafter see, there has sometimes been a complete transposition of the ordinary characters proper to each sex; the females having become the more eager in courtship, the males remaining comparatively passive, but apparently selecting, as we may infer from the results, the more attractive females. Certain female birds have thus been rendered more highly coloured or otherwise ornamented, as well as more powerful and pugnacious than the males, these characters being transmitted to the female offspring alone.

It may be suggested that in some cases a double process of selection has been carried on; the males having selected the more attractive females, and the latter the more attractive males. This process however, though it might lead to the modification of both sexes, 277would not make the one sex different from the other, unless indeed their taste for the beautiful differed; but this is a supposition too improbable in the case of any animal, excepting man, to be worth considering. There are, however, many animals, in which the sexes resemble each other, both being furnished with the same ornaments, which analogy would lead us to attribute to the agency of sexual selection. In such cases it may be suggested with more plausibility, that there has been a double or mutual process of sexual selection; the more vigorous and precocious females having selected the more attractive and vigorous males, the latter having rejected all except the more attractive females. But from what we know of the habits of animals, this view is hardly probable, the male being generally eager to pair with any female. It is more probable that the ornaments common to both sexes were acquired by one sex, generally the male, and then transmitted to the offspring of both sexes. If, indeed, during a lengthened period the males of any species were greatly to exceed the females in number, and then during another lengthened period under different conditions the reverse were to occur, a double, but not simultaneous, process of sexual selection might easily be carried on, by which the two sexes might be rendered widely different.

We shall hereafter see that many animals exist, of which neither sex is brilliantly coloured or provided with special ornaments, and yet the members of both sexes or of one alone have probably been modified through sexual selection. The absence of bright tints or other ornaments may be the result of variations of the right kind never having occurred, or of the animals themselves preferring simple colours, such as plain black or white. Obscure colours have often been acquired through natural selection for the sake of protection, and278the acquirement through sexual selection of conspicuous colours, may have been checked from the danger thus incurred. But in other cases the males have probably struggled together during long ages, through brute force, or by the display of their charms, or by both means combined, and yet no effect will have been produced unless a larger number of offspring were left by the more successful males to inherit their superiority, than by the less successful males; and this, as previously shewn, depends on various complex contingencies.

Sexual selection acts in a less rigorous manner than natural selection. The latter produces its effects by the life or death at all ages of the more or less successful individuals. Death, indeed, not rarely ensues from the conflicts of rival males. But generally the less successful male merely fails to obtain a female, or obtains later in the season a retarded and less vigorous female, or, if polygamous, obtains fewer females; so that they leave fewer, or less vigorous, or no offspring. In regard to structures acquired through ordinary or natural selection, there is in most cases, as long as the conditions of life remain the same, a limit to the amount of advantageous modification in relation to certain special ends; but in regard to structures adapted to make one male victorious over another, either in fighting or in charming the female, there is no definite limit to the amount of advantageous modification; so that as long as the proper variations arise the work of sexual selection will go on. This circumstance may partly account for the frequent and extraordinary amount of variability presented by secondary sexual characters. Nevertheless, natural selection will determine that characters of this kind shall not be acquired by the victorious males, which would be injurious to them in any high degree, either by expending too much of their vital powers, or279 by exposing them to any great danger. The development, however, of certain structures—of the horns, for instance, in certain stags—has been carried to a wonderful extreme; and in some instances to an extreme which, as far as the general conditions of life are concerned, must be slightly injurious to the male. From this fact we learn that the advantages which favoured males have derived from conquering other males in battle or courtship, and thus leaving a numerous progeny, have been in the long run greater than those derived from rather more perfect adaptation to the external conditions of life. We shall further see, and this could never have been anticipated, that the power to charm the female has been in some few instances more important than the power to conquer other males in battle.

LAWS OF INHERITANCE.

In order to understand how sexual selection has acted, and in the course of ages has produced conspicuous results with many animals of many classes, it is necessary to bear in mind the laws of inheritance, as far as they are known. Two distinct elements are included under the term "inheritance," namely the transmission and the development of characters; but as these generally go together, the distinction is often overlooked. We see this distinction in those characters which are transmitted through the early years of life, but are developed only at maturity or during old age. We see the same distinction more clearly with secondary sexual characters, for these are transmitted through both sexes, though developed in one alone. That they are present in both sexes, is manifest when two species, having strongly-marked sexual characters, are crossed, for each transmits the characters proper to280 its own male and female sex to the hybrid offspring of both sexes. The same fact is likewise manifest, when characters proper to the male are occasionally developed in the female when she grows old or becomes diseased; and so conversely with the male. Again, characters occasionally appear, as if transferred from the male to the female, as when, in certain breeds of the fowl, spurs regularly appear in the young and healthy females; but in truth they are simply developed in the female; for in every breed each detail in the structure of the spur is transmitted through the female to her male offspring. In all cases of reversion, characters are transmitted through two, three, or many generations, and are then under certain unknown favourable conditions developed. This important distinction between transmission and development will be easiest kept in mind by the aid of the hypothesis of pangenesis, whether or not it be accepted as true. According to this hypothesis, every unit or cell of the body throws off gemmules or undeveloped atoms, which are transmitted to the offspring of both sexes, and are multiplied by self-division. They may remain undeveloped during the early years of life or during successive generations; their development into units or cells, like those from which they were derived, depending on their affinity for, and union with, other units or cells previously developed in the due order of growth.

111

Inheritance at Corresponding Periods of Life.—This tendency is well established. If a new character appears in an animal whilst young, whether it endures throughout life or lasts only for a time, it will reappear, as a general rule, at the same age and in the same manner in the offspring. If, on the other hand, a new character appears at maturity, or even during old age, it tends281 to reappear in the offspring at the same advanced age. When deviations from this rule occur, the transmitted characters much oftener appear before than after the corresponding age. As I have discussed this subject at sufficient length in another work,354 I will here merely give two or three instances, for the sake of recalling the subject to the reader's mind. In several breeds of the Fowl, the chickens whilst covered with down, in their first true plumage, and in their adult plumage, differ greatly from each other, as well as from their common parent-form, the Gallus bankiva; and these characters are faithfully transmitted by each breed to their offspring at the corresponding period of life. For instance, the chickens of spangled Hamburghs, whilst covered with down, have a few dark spots on the head and rump, but are not longitudinally striped, as in many other breeds; in their first true plumage, "they are beautifully pencilled," that is each feather is transversely marked by numerous dark bars; but in their second plumage the feathers all become spangled or tipped with a dark round spot.355 Hence in this breed variations have occurred and have been transmitted at three distinct periods of life. The Pigeon offers a more remarkable case, because the aboriginal parent-species does not undergo with advancing age any change of plumage, excepting that at maturity the breast becomes more iridescent; yet there are breeds which do not acquire their characteristic colours until they 282have moulted two, three, or four times; and these modifications of plumage are regularly transmitted.

Inheritance at Corresponding Seasons of the Year.—With animals in a state of nature innumerable instances occur of characters periodically appearing at different seasons. We see this with the horns of the stag, and with the fur of arctic animals which becomes thick and white during the winter. Numerous birds acquire bright colours and other decorations during the breeding-season alone. I can throw but little light on this form of inheritance from facts observed under domestication. Pallas states,356 that in Siberia domestic cattle and horses periodically become lighter-coloured during the winter; and I have observed a similar marked change of colour in certain ponies in England. Although I do not know that this tendency to assume a differently coloured coat during different seasons of the year is transmitted, yet it probably is so, as all shades of colour are strongly inherited by the horse. Nor is this form of inheritance, as limited by season, more remarkable than inheritance as limited by age or sex.

Inheritance as Limited by Sex.—The equal transmission of characters to both, sexes is the commonest form of inheritance, at least with those animals which do not present strongly-marked sexual differences, and indeed with many of these. But characters are not rarely transferred exclusively to that sex, in which they first appeared. Ample evidence on this head has been advanced in my work on Variation under Domestica283tion; but a few instances may here be given. There are breeds of the sheep and goat, in which the horns of the male differ greatly in shape from those of the female; and these differences, acquired under domestication, are regularly transmitted to the same sex. With tortoise-shell cats the females alone, as a general rule, are thus coloured, the males being rusty-red. With most breeds of the fowl, the characters proper to each sex are transmitted to the same sex alone. So general is this form of transmission that it is an anomaly when we see in certain breeds variations transmitted equally to both sexes. There are also certain sub-breeds of the fowl in which the males can hardly be distinguished from each other, whilst the females differ considerably in colour. With the pigeon the sexes of the parent-species do not differ in any external character; nevertheless in certain domesticated breeds the male is differently coloured from the female.357 The wattle in the English Carrier pigeon and the crop in the Pouter are more highly developed in the male than in the female; and although these characters have been gained through long-continued selection by man, the difference between the two sexes is wholly due to the form of inheritance which has prevailed; for it has arisen, not from, but rather in opposition to, the wishes of the breeder.

Most of our domestic races have been formed by the accumulation of many slight variations; and as some of the successive steps have been transmitted to one sex alone, and some to both sexes, we find in the different breeds of the same species all gradations between great sexual dissimilarity and complete similarity. In284stances have already been given with the breeds of the fowl and pigeon; and under nature analogous cases are of frequent occurrence. With animals under domestication, but whether under nature I will not venture to say, one sex may lose characters proper to it, and may thus come to resemble to a certain extent the opposite sex; for instance, the males of some breeds of the fowl have lost their masculine plumes and hackles. On the other hand the differences between the sexes may be increased under domestication, as with merino sheep, in which the ewes have lost their horns. Again, characters proper to one sex may suddenly appear in the other sex; as with those sub-breeds of the fowl in which the hens whilst young acquire spurs; or, as in certain Polish sub-breeds, in which the females, as there is reason to believe, originally acquired a crest, and subsequently transferred it to the males. All these cases are intelligible on the hypothesis of pangenesis; for they depend on the gemmules of certain units of the body, although present in both sexes, becoming through the influence of domestication dormant in the one sex; or if naturally dormant, becoming developed.

There is one difficult question which it will be convenient to defer to a future chapter; namely, whether a character at first developed in both sexes, can be rendered through selection limited in its development to one sex alone. If, for instance, a breeder observed that some of his pigeons (in which species characters are usually transferred in an equal degree to both sexes) varied into pale blue; could he by long-continued selection make a breed, in which the males alone should be of this tint, whilst the females remained unchanged? I will here only say, that this, though perhaps not impossible, would be extremely difficult; for the natural result of breeding from the pale-blue males would be285 to change his whole stock, including both sexes, into this tint. If, however, variations of the desired tint appeared, which were from the first limited in their development to the male sex, there would not be the least difficulty in making a breed characterised by the two sexes being of a different colour, as indeed has been effected with a Belgian breed, in which the males alone are streaked with black. In a similar manner, if any variation appeared in a female pigeon, which was from the first sexually limited in its development, it would be easy to make a breed with the females alone thus characterised; but if the variation was not thus originally limited, the process would be extremely difficult, perhaps impossible.

On the Relation between the period of Development of a Character and its transmission to one sex or to both sexes.—Why certain characters should be inherited by both sexes, and other characters by one sex alone, namely by that sex in which the character first appeared, is in most cases quite unknown. We cannot even conjecture why with certain sub-breeds of the pigeon, black striæ, though transmitted through the female, should be developed in the male alone, whilst every other character is equally transferred to both sexes. Why, again, with cats, the tortoise-shell colour should, with rare exceptions, be developed in the female alone.

The very same characters, such as deficient or supernumerary digits, colour-blindness, &c., may with mankind be inherited by the males alone of one family, and in another family by the females alone, though in both cases transmitted through the opposite as well as the same sex.358 Although we are thus ignorant, two rules often hold good, namely 286that variations which, first appear in either sex at a late period of life, tend to be developed in the same sex alone; whilst variations which first appear early in life in either sex tend to be developed in both sexes. I am, however, far from supposing that this is the sole determining cause. As I have not elsewhere discussed this subject, and as it has an important bearing on sexual selection, I must here enter into lengthy and somewhat intricate details.

It is in itself probable that any character appearing at an early age would tend to be inherited equally by both sexes, for the sexes do not differ much in constitution, before the power of reproduction is gained. On the other hand, after this power has been gained and the sexes have come to differ in constitution, the gemmules (if I may again use the language of pangenesis) which are cast off from each varying part in the one sex would be much more likely to possess the proper affinities for uniting with the tissues of the same sex, and thus becoming developed, than with those of the opposite sex.

I was first led to infer that a relation of this kind exists, from the fact that whenever and in whatever manner the adult male has come to differ from the adult female, he differs in the same manner from the young of both sexes. The generality of this fact is quite remarkable: it holds good with almost all mammals, birds, amphibians, and fishes; also with many crustaceans, spiders and some few insects, namely certain orthoptera and libellulæ. In all these cases the variations, through the accumulation of which the male acquired his proper masculine characters, must have occurred at a somewhat late period of life; otherwise the young males would have been similarly characterised; and conformably with our rule, they are transmitted to287 and developed in the adult males alone. When, on the other hand, the adult male closely resembles the young of both sexes (these, with rare exceptions, being alike), he generally resembles the adult female; and in most of these cases the variations through which the young and old acquired their present characters, probably occurred in conformity with our rule during youth. But there is here room for doubt, as characters are sometimes transferred to the offspring at an earlier age than that at which they first appeared in the parents, so that the parents may have varied when adult, and have transferred their characters to their offspring whilst young. There are, moreover, many animals, in which the two sexes closely resemble each other, and yet both differ from their young; and here the characters of the adults must have been acquired late in life; nevertheless, these characters in apparent contradiction to our rule, are transferred to both sexes. We must not, however, overlook the possibility or even probability of successive variations of the same nature sometimes occurring, under exposure to similar conditions, simultaneously in both sexes at a rather late period of life; and in this case the variations would be transferred to the offspring of both sexes at a corresponding late age; and there would be no real contradiction to our rule of the variations which occur late in life being transferred exclusively to the sex in which they first appeared. This latter rule seems to hold true more generally than the second rule, namely, that variations which occur in either sex early in life tend to be transferred to both sexes. As it was obviously impossible even to estimate in how large a number of cases throughout the animal kingdom these two propositions hold good, it occurred to me to investigate some striking or crucial instances, and to rely on the result.

288An excellent case for investigation is afforded by the Deer Family. In all the species, excepting one, the horns are developed in the male alone, though certainly transmitted through the female, and capable of occasional abnormal development in her. In the reindeer, on the other hand, the female is provided with horns; so that in this species, the horns ought, according to our rule, to appear early in life, long before the two sexes had arrived at maturity and had come to differ much in constitution. In all the other species of deer the horns ought to appear later in life, leading to their development in that sex alone, in which they first appeared in the progenitor of the whole Family. Now in seven species, belonging to distinct sections of the family and inhabiting different regions, in which the stags alone bear horns, I find that the horns first appear at periods varying from nine months after birth in the roebuck to ten or twelve or even more months in the stags of the six other larger species.359 But with the reindeer the case is widely different, for as I hear from Prof. Nilsson, who kindly made special enquiries for me in Lapland, the horns appear in the young animals within four or five weeks after birth, and at the same time in both sexes. So that here we have a structure, developed at a most unusually early age in one species of the family, and common to both sexes in this one species.

In several kinds of antelopes the males alone are 289provided with horns, whilst in the greater number both sexes have horns. With respect to the period of development, Mr. Blyth informs me that there lived at one time in the Zoological Gardens a young koodoo (Ant. strepsiceros), in which species the males alone are horned, and the young of a closely-allied species, viz. the eland (Ant. oreas), in which both sexes are horned. Now in strict conformity with our rule, in the young male koodoo, although arrived at the age of ten months, the horns were remarkably small considering the size ultimately attained by them: whilst in the young male eland, although only three months old, the horns were already very much larger than in the koodoo. It is also worth notice that in the prong-horned antelope,360 in which species the horns, though present in both sexes, are almost rudimentary in the female, they do not appear until about five or six months after birth. With sheep, goats, and cattle, in which the horns are well developed in both sexes, though not quite equal in size, they can be felt, or even seen, at birth or soon afterwards.361 Our rule, however, fails in regard to some breeds of sheep, for instance merinos, in which the rams alone are horned; for I cannot find on enquiry,362 that 290the horns are developed later in life in this breed than in ordinary sheep in which both sexes are horned. But with domesticated sheep the presence or absence of horns is not a firmly fixed character; a certain proportion of the merino ewes bearing small horns, and some of the rams being hornless; whilst with ordinary sheep hornless ewes are occasionally produced.

In most of the species of the splendid family of the Pheasants, the males differ conspicuously from the females, and they acquire their ornaments at a rather late period of life. The eared pheasant (Crossoptilon auritum), however, offers a remarkable exception, for both sexes possess the fine caudal plumes, the large ear-tufts and the crimson velvet about the head; and I find on enquiry in the Zoological Gardens that all these characters, in accordance with our rule, appear very early in life. The adult male can, however, be distinguished from the adult female by one character, namely by the presence of spurs; and conformably with our rule, these do not begin to be developed, as I am assured by Mr. Bartlett, before the age of six months, and even at this age, can hardly be distinguished in the two sexes.363 The male and female Peacock differ con291spicuously from each other in almost every part of their plumage, except in the elegant head-crest, which is common to both sexes; and this is developed very early in life, long before the other ornaments which are confined to the male. The wild-duck offers an analogous case, for the beautiful green speculum on the wings is common to both sexes, though duller and somewhat smaller in the female, and it is developed early in life, whilst the curled tail-feathers and other ornaments peculiar to the male are developed later.364 Between such extreme cases of close sexual resemblance and wide dissimilarity, as those of the Crossoptilon and peacock, many intermediate ones could be given, in which the characters follow in their order of development our two rules.

As most insects emerge from their pupal state in a mature condition, it is doubtful whether the period of development determines the transference of their characters to one or both sexes. But we do not know that the coloured scales, for instance, in two species of butterflies, in one of which the sexes differ in colour, whilst in the other they are alike, are developed at the same relative age in the cocoon. Nor do we know whether all the scales are simultaneously developed on the wings 292of the same species of butterfly, in which certain coloured marks are confined to one sex, whilst other marks are common to both sexes. A difference of this kind in the period of development is not so improbable as it may at first appear; for with the Orthoptera, which assume their adult state, not by a single metamorphosis, but by a succession of moults, the young males of some species at first resemble the females, and acquire their distinctive masculine characters only during a later moult. Strictly analogous cases occur during the successive moults of certain male crustaceans.

We have as yet only considered the transference of characters, relatively to their period of development, with species in a natural state; we will now turn to domesticated animals; first touching on monstrosities and diseases. The presence of supernumerary digits, and the absence of certain phalanges, must be determined at an early embryonic period—the tendency to profuse bleeding is at least congenital, as is probably colour-blindness—yet these peculiarities, and other similar ones, are often limited in their transmission to one sex; so that the rule that characters which are developed at an early period tend to be transmitted to both sexes, here wholly fails. But this rule, as before remarked, does not appear to be nearly so generally true as the converse proposition, namely, that characters which appear late in life in one sex are transmitted exclusively to the same sex. From the fact of the above abnormal peculiarities becoming attached to one sex. long before the sexual functions are active, we may infer that there must be a difference of some kind between the sexes at an extremely early age. With respect to sexually-limited diseases, we know too little of the period at which they originate, to draw any fair conclusion. Gout, however, seems to fall under293 our rule; for it is generally caused by intemperance after early youth, and is transmitted from the father to his sons in a much more marked manner than to his daughters.

In the various domestic breeds of sheep, goats, and cattle, the males differ from their respective females in the shape or development of their horns, forehead, mane, dewlap, tail, and hump on the shoulders; and these peculiarities, in accordance with our rule, are not fully developed until rather late in life. With dogs, the sexes do not differ, except that in certain breeds, especially in the Scotch deer-hound, the male is much larger and heavier than the female; and as we shall see in a future chapter, the male goes on increasing in size to an unusually late period of life, which will account, according to our rule, for his increased size being transmitted to his male offspring alone. On the other hand, the tortoise-shell colour of the hair, which is confined to female cats, is quite distinct at birth, and this case violates our rule. There is a breed of pigeons in which the males alone are streaked with black, and the streaks can be detected even in the nestlings; but they become more conspicuous at each successive moult, so that this case partly opposes and partly supports the rule. With the

English Carrier and Pouter pigeon the full development of the wattle and the crop occurs rather late in life, and these characters, conformably with our rule, are transmitted in full perfection to the males alone. The following cases perhaps come within the class previously alluded to, in which the two sexes have varied in the same manner at a rather late period of life, and have consequently transferred their new characters to both sexes at a corresponding late period; and if so, such cases are not opposed to our rule. Thus there are sub-breeds of the pigeon, described by Neumeis294ter,365 both sexes of which change colour after moulting twice or thrice, as does likewise the Almond Tumbler; nevertheless these changes, though occurring rather late in life, are common to both sexes. One variety of the Canary-bird, namely the London Prize, offers a nearly analogous case.

With the breeds of the Fowl the inheritance of various characters by one sex or by both sexes, seems generally determined by the period at which such characters are developed. Thus in all the many breeds in which the adult male differs greatly in colour from the female and from the adult male parent-species, he differs from the young male, so that the newly acquired characters must have appeared at a rather late period of life. On the other hand with most of the breeds in which the two sexes resemble each other, the young are coloured in nearly the same manner as their parents, and this renders it probable that their colours first appeared early in life. We have instances of this fact in all black and white breeds, in which the young and old of both sexes are alike; nor can it be maintained that there is something peculiar in a black or white plumage, leading to its transference to both sexes; for the males alone of many natural species are either black or white, the females being very differently coloured. With the so-called Cuckoo sub-breeds of the fowl, in which the feathers are transversely pencilled with dark stripes, both sexes and the chickens are coloured in nearly the same manner. The laced plumage of the Sebright bantam is the same in both sexes, and in the chickens the feathers are tipped with black, which makes a near approach to lacing. Spangled Hamburghs, however, offer a partial exception, 295for the two sexes, though not quite alike, resemble each other more closely than do the sexes of the aboriginal parent-species, yet they acquire their characteristic plumage late in life, for the chickens are distinctly pencilled. Turning to other characters besides colour: the males alone of the wild parent-species and of most domestic breeds possess a fairly well developed comb, but in the young of the Spanish fowl it is largely developed at a very early age, and apparently in consequence of this it is of unusual size in the adult females. In the Game breeds pugnacity is developed at a wonderfully early age, of which curious proofs could be given; and this character is transmitted to both sexes, so that the hens, from their extreme pugnacity, are now generally exhibited in separate pens. With the Polish breeds the bony protuberance of the skull which supports the crest is partially developed even before the chickens are hatched, and the crest itself soon begins to grow, though at first feebly;366 and in this breed a great bony protuberance and an immense crest characterise the adults of both sexes.

Finally, from what we have now seen of the relation which exists in many natural species and domesticated races, between the period of the development of their characters and the manner of their transmission—for example the striking fact of the early growth of the horns in the reindeer, in which both sexes have horns, in comparison with their much later growth in the other species in which the male alone bears horns—we may conclude that one cause, though not the sole 296cause, of characters being exclusively inherited by one sex, is their development at a late age. And secondly, that one, though apparently a less efficient, cause of characters being inherited by both sexes is their development at an early age, whilst the sexes differ but little in constitution. It appears, however, that some difference must exist between the sexes even during an early embryonic period, for characters developed at this age not rarely become attached to one sex.

Summary and concluding remarks.—From the foregoing discussion on the various laws of inheritance, we learn that characters often or even generally tend to become developed in the same sex, at the same age, and periodically at the same season of the year, in which they first appeared in the parents. But these laws, from unknown causes, are very liable to change. Hence the successive steps in the modification of a species might readily be transmitted in different ways; some of the steps being transmitted to one sex, and some to both; some to the offspring at one age, and some at all ages. Not only are the laws of inheritance extremely complex, but so are the causes which induce and govern variability. The variations thus caused are preserved and accumulated by sexual selection, which is in itself an extremely complex affair, depending, as it does, on ardour in love, courage, and the rivalry of the males, and on the powers of perception, taste, and will of the female. Sexual selection will also be dominated by natural selection for the general welfare of the species. Hence the manner in which the individuals of either sex or of both sexes are affected through sexual selection cannot fail to be complex in the highest degree.

When variations occur late in life in one sex, and are297 transmitted to the same sex at the same age, the other sex and the young are necessarily left unmodified. When they occur late in life, but are transmitted to both sexes at the same age, the young alone are left unmodified. Variations, however, may occur at any period of life in one sex or in both, and be transmitted to both sexes at all ages, and then all the individuals of the species will be similarly modified. In the following chapters it will be seen that all these cases frequently occur under nature.

Sexual selection can never act on any animal whilst young, before the age for reproduction has arrived. From the great eagerness of the male it has generally acted on this sex and not on the females. The males have thus become provided with weapons for fighting with their rivals, or with organs for discovering and securely holding the female, or for exciting and charming her. When the sexes differ in these respects, it is also, as we have seen, an extremely general law that the adult male differs more or less from the young male; and we may conclude from this fact that the successive variations, by which the adult male became modified, cannot have occurred much before the age for reproduction. How then are we to account for this general and remarkable coincidence between the period of variability and that of sexual selection,—principles which are quite independent of each other? I think we can see the cause: it is not that the males have never varied at an early age, but that such variations have commonly been lost, whilst those occurring at a later age have been preserved.

All animals produce more offspring than can survive to maturity; and we have every reason to believe that death falls heavily on the weak and inexperienced young. If then a certain proportion of the offspring298 were to vary at birth or soon afterwards, in some manner which at this age was of no service to them, the chance of the preservation of such variations would be small. We have good evidence under domestication how soon variations of all kinds are lost, if not selected. But variations which occurred at or near maturity, and which were of immediate service to either sex, would probably be preserved; as would similar variations occurring at an earlier period in any individuals which happened to survive. As this principle has an important bearing on sexual selection, it may be advisable to give an imaginary illustration. We will take a pair of animals, neither very fertile nor the reverse, and assume that after arriving at maturity they live on an average for five years, producing each year five young. They would thus produce 25 offspring; and it would not, I think, be an unfair estimate to assume that 18 or 20 out of the 25 would perish before maturity, whilst still young and inexperienced; the remaining seven or five sufficing to keep up the stock of mature individuals. If so, we can see that variations which occurred during youth, for instance in brightness, and which were not of the least service to the young, would run a good chance of being utterly lost. Whilst similar variations, which occurring at or near maturity in the comparatively few individuals surviving to this age, and which immediately gave an advantage to certain males, by rendering them more attractive to the females, would be likely to be preserved. No doubt some of the variations in brightness which occurred at an earlier age would by chance be preserved, and eventually give to the male the same advantage as those which appeared later; and this will account for the young males commonly partaking to a certain extent (as may be observed with many birds) of the bright colours of their299 adult male parents. If only a few of the successive variations in brightness were to occur at a late age, the adult male would be only a little brighter than the young male; and such cases are common.

In this illustration I have assumed that the young varied in a manner which was of no service to them; but many characters proper to the adult male would be actually injurious to the young,—as bright colours from making them conspicuous, or horns of large size from expending much vital force. Such variations in the young would promptly be eliminated through natural selection. With the adult and experienced males, on the other hand, the advantage thus derived in their rivalry with other males would often more than counterbalance exposure to some degree of danger. Thus we can understand how it is that variations which must originally have appeared rather late in life have alone or in chief part been preserved for the development of secondary sexual characters; and the remarkable coincidence between the periods of variability and of sexual selection is intelligible.

As variations which give to the male an advantage in lighting with other males, or in finding, securing, or charming the female, would be of no use to the female, they will not have been preserved in this sex either during youth or maturity. Consequently such variations would be extremely liable to be lost; and the female, as far as these characters are concerned, would be left unmodified, excepting in so far as she may have received them by transference from the male. No doubt if the female varied and transferred serviceable characters to her male offspring, these would be favoured through sexual selection; and then both sexes would thus far be modified in the same manner. But I shall hereafter have to recur to these more intricate contingencies.

300In the following chapters, I shall treat of the secondary sexual characters in animals of all classes, and shall endeavour in each case to apply the principles explained in the present chapter. The lowest classes will detain us for a very short time, but the higher animals, especially birds, must be treated at considerable length. It should be borne in mind that for reasons already assigned, I intend to give only a few illustrative instances of the innumerable structures by the aid of which the male finds the female, or, when found, holds her. On the other hand, all structures and instincts by which the male conquers other males, and by which he allures or excites the female, will be fully discussed, as these are in many ways the most interesting.

Supplement on the proportional numbers of the two sexes in animals belonging to various classes.

As no one, as far as I can discover, has paid attention to the relative numbers of the two sexes throughout the animal kingdom, I will here give such materials as I have been able to collect, although they are extremely imperfect. They consist in only a few instances of actual enumeration, and the numbers are not very large. As the proportions are known with certainty on a large scale in the case of man alone, I will first give them, as a standard of comparison.

Man.—In England during ten years (from 1857 to 1866) 707,120 children on an annual average have been born alive, in the proportion of 104.5 males to 100 females. But in 1857 the male births throughout England were as 105.2, and in 1865 as 104.0 to 100. Looking to separate districts, in Buckinghamshire (where on an average 5000 children are annually born)301 the mean proportion of male to female births, during the whole period of the above ten years, was as 102.8 to 100; whilst in N. Wales (where the average annual births are 12,873) it was as high as 106.2 to 100. Taking a still smaller district, viz., Rutlandshire (where the annual births average only 739), in 1864 the male births were as 114.6, and in 1862 as 97.0 to 100; but even in this small district the average of the 7385 births during the whole ten years was as 104.5 to 100; that is in the same ratio as throughout England.367 The proportions are sometimes slightly disturbed by unknown causes; thus Prof. Faye states "that in some districts of Norway there has been during a decennial period a steady deficiency of boys, whilst in others the opposite condition has existed." In France during forty-four years the male to the female births have been as 106.2 to 100; but during this period it has occurred five times in one department, and six times in another, that the female births have exceeded the males. In Russia the average proportion is as high as 108.9 to 100.368 It is a singular fact that with Jews the proportion of male births is decidedly larger than with Christians: thus in Prussia the proportion is as 113, in Breslau as 114, and in Livonia as 120 to 100; the Christian births in these countries being the same as usual, for instance, in Livonia as 104 to 100.369 It is a still more singular fact that in different nations, under different conditions and climates, in Naples, Prussia, Westphalia, France and England, the 302excess of male over female births is less when they are illegitimate than when legitimate.370

In various parts of Europe, according to Prof. Faye and other authors, "a still greater preponderance of males would be met with, if death struck both sexes in equal proportion in the womb and during birth. But the fact is, that for every 100 still-born females, we have in several countries from 134.6 to 144.9 still-born males." Moreover during the first four or five years of life more male children die than females; "for example in England, during the first year, 126 boys die for every 100 girls,—a proportion which in France is still more unfavourable."371 As a consequence of this excess in the death-rate of male children, and of the exposure of men when adult to various dangers, and of their tendency to emigrate, the females in all old-settled countries, where statistical records have been kept,372 are found to preponderate considerably over the males.

It has often been supposed that the relative ages of the parents determine the sex of the offspring; and Prof. Leuckart373 has advanced what he considers 303sufficient evidence, with respect to man and certain domesticated animals, to shew that this is one important factor in the result. So again the period of impregnation has been thought to be the efficient cause; but recent observations discountenance this belief. Again, with mankind polygamy has been supposed to lead to the birth of a greater proportion of female infants; but Dr. J. Campbell374 carefully attended to this subject in the harems of Siam, and he concludes that the proportion of male to female births is the same as from monogamous unions. Hardly any animal has been rendered so highly polygamous as our English race-horses, and we shall immediately see that their male and female offspring are almost exactly equal in number.

Horses.—Mr. Tegetmeier has been so kind as to tabulate for me from the 'Racing Calendar' the births of race-horses during a period of twenty-one years, viz. from 1846 to 1867; 1849 being omitted, as no returns were that year published. The total births have been 25,560,<u>375</u> consisting of 12,763 males and 12,797 females, or in the proportion of 99.7 males to 100 females. As these numbers are tolerably large, and as they are drawn from all parts of England, during several years, we may with much confidence conclude that with the domestic horse, or at least with the race-horse, the two sexes are produced in almost equal numbers. The fluctuations in the proportions during successive years are closely like those which occur with mankind, when a small and thinly-populated area is considered: thus in 1856 the male horses were as 107.1, and in 1867 as only 92.6 to 100 females. In the tabulated returns the proportions vary in cycles, for the males exceeded the females during six successive years; and the females exceeded the males during two 304periods each of four years: this, however, may be accidental; at least I can detect nothing of the kind with man in the decennial table in the Registrar's Report for 1866. I may add that certain, mares, and this holds good with certain cows and with women, tend to produce more of one sex than of the other; Mr. Wright of Yeldersley House, informs me that one of his Arab mares, though put seven times to different horses, produced seven fillies.

Dogs.—During a period of twelve years, from 1857 to 1868, the births of a large number of greyhounds, throughout England, have been sent to the 'Field' newspaper; and I am again indebted to Mr. Tegetmeier for carefully tabulating the results. The recorded, births have been 6878, consisting of 3605 males and 3273 females, that is, in the proportion of 110.1 males to 100 females. The greatest fluctuations occurred in 1864, when the proportion was as 95.3 males, and in 1867, as 116.3 males to 100 females. The above average proportion of 110.1 to 100 is probably nearly correct in the case of the greyhound, but whether it would hold with other domesticated breeds is in some degree doubtful. Mr. Cupples has enquired from several great breeders of dogs, and finds that all without exception believe that females are produced in excess; he suggests that this belief may have arisen from females being less valued and the consequent disappointment producing a stronger impression on the mind.

Sheep.—The sexes of sheep are not ascertained by agriculturists until several months after birth, at the period when the males are castrated; so that the following returns do not give the proportions at birth. Moreover, I find that several great breeders in Scotland, who annually raise some thousand sheep, are firmly convinced that a larger proportion of males than of females die during the first one or two years; therefore the proportion of males would be somewhat greater at birth than at the age of castration. This is a remarkable coincidence with what occurs, as we have seen, with mankind, and both cases probably depend on some common cause. I have received returns from four gentlemen in England who have bred lowland sheep, chiefly Leicesters, during the last ten or sixteen years; they amount altogether to 8965 births, consisting of 4407 males and 4558 females; that is in the proportion of 96.7 males to 100 females. With respect to Cheviot and black-faced sheep bred in Scotland, I have received returns from six breeders, two of them on a large scale, chiefly for the years 1867-1869, but some of the returns extending back to 1862. The total number recorded amounts to 50,685, consisting of 25,071 males and 25,614 females, or in the proportion of 97.9 males to 100 females. If we take the English and Scotch returns together, the total number amounts 305to 59,650, consisting of 29,478 males and 30,172 females, or as 97·7 to 100. So that with sheep at the age of castration the females are certainly in excess of the males; but whether this would hold good at birth is doubtful, owing to the greater liability in the males to early death.376

Of Cattle I have received returns from nine gentlemen of 982 births, too few to be trusted; these consisted of 477 bull-calves and 505 cow-calves; i.e. in the proportion of 94·4 males to 100 females. The Rev. W. D. Fox informs me that in 1867 out of 34 calves born on a farm in Derbyshire only one was a bull. Mr. Harrison Weir writes to me that he has enquired from several breeders of Pigs, and most of them estimate the male to the female births as about 7 to 6. This same gentleman has bred Rabbits for many years, and has noticed that a far greater number of bucks are produced than does.

Of mammalia in a state of nature I have been able to learn very little. In regard to the common rat, I have received conflicting statements. Mr. R. Elliot of Laighwood, informs me that a rat-catcher assured him that he had always found the males in great excess, even with the young in the nest. In consequence of this, Mr. Elliot himself subsequently examined some hundred old ones, and found the statement true. Mr. F. Buckland has bred a large number of white rats, and he also believes that the males greatly exceed the females. In regard to Moles, it is said that "the males are much more numerous than the females;"377 and as the catching of these animals is a special occupation, the statement may perhaps be trusted. Sir A. Smith, in describing an antelope of S. Africa378 (Kobus ellipsiprymnus), remarks, that in the herds of this and other species, the males are few in number compared with the females: the natives believe that they are born in this proportion; others believe that the younger males are expelled from the herds, and Sir A. Smith says, that though he has himself never seen herds consisting of young males alone, others affirm that this does occur. It appears probable that the young males when expelled from the herd, would be likely to fell a prey to the many beasts of prey of the country.

BIRDS.

With respect to the Fowl, I have received only one account, namely, that out of 1001 chickens of a highly-bred stock of Cochins, reared during eight years by Mr. Stretch, 487 proved males and 514 females: i.e. as 94.7 to 100. In regard to domestic pigeons there is good evidence that the males are produced in excess, or that their lives are longer; for these birds invariably pair, and single males, as Mr. Tegetmeier informs me, can always be purchased cheaper than females. Usually the two birds reared from the two eggs laid in the same nest consist of a male and female; but Mr. Harrison Weir, who has been so large a breeder, says that he has often bred two cocks from the same nest, and seldom two hens; moreover the hen is generally the weaker of the two, and more liable to perish.

With respect to birds in a state of nature, Mr. Gould and others379 are convinced that the males are generally the more numerous; and as the young males of many species resemble the females, the latter would naturally appear to be the most numerous. Large numbers of pheasants are reared by Mr. Baker of Leadenhall from eggs laid by wild birds, and he informs Mr. Jenner Weir that four or five males to one female are generally produced. An experienced observer remarks380 that in Scandinavia the broods of the capercailzie and black-cock contain more males than females; and that with the Dal-ripa (a kind of ptarmigan) more males than females attend the leks or places of courtship; but this latter circumstance is accounted for by some observers by a greater number of hen birds being killed by vermin. From various facts given by White of Selbourne,381 it seems clear that the males of the partridge must be in considerable excess in the south of England; and I have been assured that this is the case in Scotland. Mr. Weir on enquiring from the dealers who receive at certain seasons large numbers of ruffs (Machetes pugnax) was told that the males are much the most numerous. This same naturalist has also enquired for me from the bird-catchers, who annually catch an astonishing number of various small species alive for the London market, and he was unhesitatingly answered by an old and trustworthy man, that with the chaffinch the males are in large excess; he thought as high as 2 males to 3071 female, or at least as high as 5 to 3.382 The males of the blackbird, he likewise maintained, were by far the most numerous, whether caught by traps or by netting at night. These statements may apparently be trusted, because the same man said that the sexes are about equal with the lark, the twite (Linaria montana), and goldfinch. On the other hand he is certain that with the common linnet, the females preponderate greatly, but unequally during different years; during some years he has found the females to the males as four to one. It should, however, be borne in mind, that the chief season for catching birds does not begin till September, so that with some species partial migrations may have begun, and the flocks at this period often consist of hens alone. Mr. Salvin paid particular attention to the sexes of the humming-birds in Central America, and he is convinced that with most of the species the males are in excess; thus one year he procured 204 specimens belonging to ten species, and these consisted of 166 males and of 38 females. With two other species the females were in excess: but the proportions apparently vary either during different seasons or in different localities; for on one occasion the males of Campylopterus hemileucurus were to the females as five to two, and on another occasion383 in exactly the reversed ratio. As bearing on this latter point, I may add, that Mr. Powys found in Corfu and Epirus the sexes of the chaffinch keeping apart, and "the females by far the most numerous;" whilst in Palestine Mr. Tristram found "the male flocks appearing greatly to exceed the female in number."384 So again with the Quiscalus major, Mr. G. Taylor385 says, that in Florida there were "very few females in proportion to the males," whilst in Honduras the proportion was the other way, the species there having the character of a polygamist.

FISH.

With Fish the proportional numbers of the sexes can be ascertained only by catching them in the adult or nearly adult state; and there 308are many difficulties in arriving at any just conclusion.386Infertile females might readily be mistaken for males, as Dr. Günther has remarked to me in regard to trout. With some species the males are believed to die soon after fertilising the ova. With many species the males are of much smaller size than the females, so that a large number of males would escape from the same net by which the females were caught. M. Carbonnier,387 who has especially attended to the natural history of the pike (Esox lucius) states that many males, owing to their small size, are devoured by the larger females; and he believes that the males of almost all fish are exposed from the same cause to greater danger than the females. Nevertheless in the few cases in which the proportional numbers have been actually observed, the males appear to be largely in excess. Thus Mr. R. Buist, the superintendent of the Stormontfield experiments, says that in 1865, out of 70 salmon first landed for the purpose of obtaining the ova, upwards of 60 were males. In 1867 he again "calls attention to the vast disproportion of the males to the females. We had at the outset at least ten males to one female." Afterwards sufficient females for obtaining ova were procured. He adds, "from the great proportion of the males, they are constantly fighting and tearing each other on the spawning-beds."388 This disproportion, no doubt, can be accounted for in part, but whether wholly is very doubtful, by the males ascending the rivers before the females. Mr. F. Buckland remarks in regard to trout, that "it is a curious fact that the males preponderate very largely in number over the females. It invariably happens that when the first rush of fish is made to the net, there will be at least seven or eight males to one female found captive. I cannot quite account for this; either the males are more numerous than the females, or the latter seek safety by concealment rather than flight." He then adds, that by carefully searching the banks, sufficient females for obtaining ova can be found.389 Mr. H. Lee informs me that out of 212 trout, taken for this purpose in Lord Portsmouth's park, 150 were males and 62 females.

With the Cyprinidæ the males likewise seem to be in excess; but several members of this Family, viz., the carp, tench, bream and minnow, appear regularly to follow the practice, rare in the309animal kingdom, of polyandry; for the female whilst spawning is always attended by two males, one on each side, and in the case of the bream by three or four males. This fact is so well known, that it is always recommended to stock a pond with two male tenches to one female, or at least with three males to two females. With the minnow, an excellent observer states, that on the spawning-beds the males are ten times as numerous as the females; when a female comes amongst the males, "she is immediately pressed closely by a male on each side; and when they have been in that situation for a time, are superseded by other two males."390

INSECTS.

In this class, the Lepidoptera alone afford the means of judging of the proportional numbers of the sexes; for they have been collected with special care by many good observers, and have been largely bred from the egg or caterpillar state. I had hoped that some breeders of silk-moths might have kept an exact record, but after writing to France and Italy, and consulting various treatises, I cannot find that this has ever been done. The general opinion appears to be that the sexes are nearly equal, but in Italy as I hear from Professor Canestrini, many breeders are convinced that the females are produced in excess. The same naturalist, however, informs me, that in the two yearly broods of the Ailanthus silk-moth (Bombyx cynthia), the males greatly preponderate in the first, whilst in the second the two sexes are nearly equal, or the females rather in excess.

In regard to Butterflies in a state of nature, several observers have been much struck by the apparently enormous preponderance of the males.391 Thus Mr. Bates,392 in speaking of the species, no less than about a hundred in number, which inhabit the Upper Amazons, says that the males are much more numerous than the females, even in the proportion of a hundred to one. In North America, Edwards, who had great experience, estimates in the genus Papilio the males to the females as four to one; and Mr. 310Walsh, who informed me of this statement, says that withP. turnus this is certainly the case. In South Africa, Mr. R. Trimen found the males in excess in 19 species;393 and in one of these, which swarms in open places, he estimated the number of males as fifty to one female. With another species, in which the males are numerous in certain localities, he collected during seven years only five females. In the island of Bourbon, M. Maillard states that the males of one species of Papilio are twenty times as numerous as the females.394 Mr. Trimen informs me that as far as he has himself seen, or heard from others, it is rare for the females of any butterfly to exceed in number the males; but this is perhaps the case with three South African species. Mr. Wallace395 states that the females of Ornithoptera crœsus, in the Malay archipelago, are more common and more easily caught than the males; but this is a rare butterfly. I may here add, that in Hyperythra, a genus of moths, Guenée says, that from four to five females are sent in collections from India for one male.

When this subject of the proportional numbers of the sexes of insects was brought before the Entomological Society,396 it was generally admitted that the males of most Lepidoptera, in the adult or imago state, are caught in greater numbers than the females; but this fact was attributed by various observers to the more retiring habits of the females, and to the males emerging earlier from the cocoon. This latter circumstance is well known to occur with most Lepidoptera, as well as with other insects. So that, as M. Personnat remarks, the males of the domesticated Bombyx yamamai, are lost at the beginning of the season, and the females at the end, from the want of mates.397 I cannot however persuade myself that these causes suffice to explain the great excess of males in the cases, above given, of butterflies which are extremely common in their native countries. Mr. Stainton, who has paid such close attention during many years to the smaller moths, informs me that when he collected them in the imago state, he thought that the males were ten times as numerous as the females, but that since he has reared them on a large scale from the caterpillar state, he is convinced that the females are the most 311numerous. Several entomologists concur in this view. Mr. Doubleday, however, and some others, take an opposite view, and are convinced that they have reared from the egg and caterpillar states a larger proportion of males than of females.

Besides the more active habits of the males, their earlier emergence from the cocoon, and their frequenting in some cases more open stations, other causes may be assigned for an apparent or real difference in the proportional numbers of the sexes of Lepidoptera, when captured in the imago state, and when reared from the egg or caterpillar state. It is believed by many breeders in Italy, as I hear from Professor Canestrini, that the female caterpillar of the silk-moth suffers more from the recent disease than the male; and Dr. Staudinger informs me that in rearing Lepidoptera more females die in the cocoon than males. With many species the female caterpillar is larger than the male, and a collector would naturally choose the finest specimens, and thus unintentionally collect a larger number of females. Three collectors have told me that this was their practice; but Dr. Wallace is sure that most collectors take all the specimens which they can find of the rarer kinds, which alone are worth the trouble of rearing. Birds when surrounded by caterpillars would probably devour the largest; and Professor Canestrini informs me that in Italy some breeders believe, though on insufficient evidence, that in the first brood of the Ailanthus silk-moth, the wasps destroy a larger number of the female than of the male caterpillars. Dr. Wallace further remarks that female caterpillars, from being larger than the males, require more time for their development and consume more food and moisture; and thus they would be exposed during a longer time to danger from ichneumons, birds, &c., and in times of scarcity would perish in greater numbers. Hence it appears quite possible that, in a state of nature, fewer female Lepidoptera may reach maturity than males; and for our special object we are concerned with the numbers at maturity, when the sexes are ready to propagate their kind.

The manner in which the males of certain moths congregate in extraordinary numbers round a single female, apparently indicates a great excess of males, though this fact may perhaps be accounted for by the earlier emergence of the males from their cocoons. Mr. Stainton informs me that from twelve to twenty males may often be seen congregated round a female Elachista rufocinerea. It is well known that if a virgin Lasiocampa quercus or Saturnia carpini be exposed in a cage, vast numbers of males collect round her, and if confined in a room will even come down the chimney to her. 312Mr. Doubleday believes that he has seen from fifty to a hundred males of both these species attracted in the course of a single day by a female under confinement. Mr. Trimen exposed in the Isle of Wight a box in which a female of the Lasiocampa had been confined on the previous day, and five males soon endeavoured to gain admittance. M. Verreaux, in Australia, having placed the female of a small Bombyx in a box in his pocket, was followed by a crowd of males, so that about 200 entered the house with him.[398]

Mr. Doubleday has called my attention to Dr. Staudinger's[399] list of Lepidoptera, which gives the prices of the males and females of 300 species or well-marked varieties of (Rhopalocera) butterflies. The prices for both sexes of the very common species are of course the same; but with 114 of the rarer species they differ; the males being in all cases, excepting one, the cheapest. On an average of the prices of the 113 species, the price of the male to that of the female is as 100 to 149; and this apparently indicates that inversely the males exceed the females in number in the same proportion. About 2000 species or varieties of moths (Heterocera) are catalogued, those with wingless females being here excluded on account of the difference in habits of the two sexes: of these 2000 species, 141 differ in price according to sex, the males of 130 being cheaper, and the males of only 11 being dearer than the females. The average price of the males of the 130 species, to that of the females, is as 100 to 143. With respect to the butterflies in this priced list, Mr. Doubleday thinks (and no man in England has had more experience), that there is nothing in the habits of the species which can account for the difference in the prices of the two sexes, and that it can be accounted for only by an excess in the numbers of the males. But I am bound to add that Dr. Staudinger himself, as he informs me, is of a different opinion. He thinks that the less active habits of the females and the earlier emergence of the males will account for his collectors securing a larger number of males than of females, and consequently for the lower prices of the former With respect to specimens reared from the caterpillar-state, Dr. Staudinger believes, as previously stated, that a greater number of females than of males die under confinement in the cocoons. He adds that with certain species one sex seems to preponderate over the other during certain years.

So that in these eight lots of cocoons and eggs, males were produced in excess. Taken together the proportion of males is as 122.7 to 100 females. But the numbers are hardly large enough to be trustworthy.

On the whole, from the above various sources of evidence, all pointing to the same direction, I infer that with most species of Lepidoptera, the males in the imago state generally exceed the females in number, whatever the proportions may be at their first emergence from the egg.

With reference to the other Orders of insects, I have been able to collect very little reliable information. With the stag-beetle (Lucanus cervus) "the males appear to be much more numerous than the females;" but when, as Cornelius remarked during 1867, an unusual number of these beetles appeared in one part of Germany, the females appeared to exceed the males as six so one. With one of the Elateridæ, the males are said to be much more numerous than the females, and "two or three are often found united with one female;"[401] so that here polyandry seems to prevail. 314With Siagonium (Staphylinidæ), in which the males are furnished with horns, "the females are far more numerous than the opposite sex." Mr. Janson stated at the Entomological Society that the females of the bark-feeding Tomicus villosus are so common as to be a plague, whilst the males are so rare as to be hardly known. In other Orders, from unknown causes, but apparently in some instances owing to parthenogenesis, the males of certain species have never been discovered or are excessively rare, as with several of the Cynipidæ.[402] In all the gall-making Cynipidæ known to Mr. Walsh, the females are four or five times as numerous as the males; and so it is, as he informs me, with the gall-making Cecidomyiæ (Diptera). With some common species of

Saw-flies (Tenthredinæ) Mr. F. Smith has reared hundreds of specimens from larvæ of all sizes, but has never reared a single male: on the other hand Curtis says,403 that with certain species (Athalia), bred by him, the males to the females were as six to one; whilst exactly the reverse occurred with the mature insects of the same species caught in the fields. With the Neuroptera, Mr. Walsh states that in many, but by no means in all, the species of the Odonatous groups (Ephemerina), there is a great overplus of males: in the genus Hetærina, also, the males are generally at least four times as numerous as the females. In certain species in the genus Gomphus the males are equally numerous, whilst in two other species, the females are twice or thrice as numerous as the males. In some European species of Psocus thousands of females may be collected without a single male, whilst with other species of the same genus both sexes are common.404 In England, Mr. MacLachlan has captured hundreds of the female Apatania muliebris, but has never seen the male; and of Boreus hyemalis only four or five males have been here seen.405 With most of these species (excepting, as I have heard, with the Tenthredinæ) there is no reason to suppose that the females are subject to parthenogenesis; and thus we see how ignorant we are on the causes of the apparent discrepancy in the proportional numbers of the two sexes.

In the other Classes of the Articulata I have been able to collect still less information. With Spiders, Mr. Blackwall, who has carefully attended to this class during many years, writes to me that the males from their more erratic habits are more commonly seen, 315and therefore appear to be the more numerous. This is actually the case with a few species; but he mentions several species in six genera, in which the females appear to be much more numerous than the males.406 The small size of the males in comparison with the females, which is sometimes carried to an extreme degree, and their widely different appearance, may account in some instances for their rarity in collections.407

Some of the lower Crustaceans are able to propagate their kind asexually, and this will account for the extreme rarity of the males. With some other forms (as with Tanais and Cypris) there is reason to believe, as Fritz Müller informs me, that the male is much shorter-lived than the female, which, supposing the two sexes to be at first equal in number, would explain the scarcity of the males. On the other hand this same naturalist has invariably taken, on the shores of Brazil, far more males than females of the Diastylidæ and of Cypridina; thus with a species in the latter genus, 63 specimens caught the same day, included 57 males; but he suggests that this preponderance may be due to some unknown difference in the habits of the two sexes. With one of the higher Brazilian crabs, namely a Gelasimus, Fritz Müller found the males to be more numerous than the females. The reverse seems to be the case, according to the large experience of Mr. C. Spence Bate, with six common British crabs, the names of which he has given me.

On the Power of Natural Selection to regulate the proportional Numbers of the Sexes, and General Fertility.—In some peculiar cases, an excess in the number of one sex over the other might be a great advantage to a species, as with the sterile females of social insects, or with those animals in which more than one male is requisite to fertilise the female, as with certain cirripedes and perhaps certain fishes. An inequality between the sexes in these cases might have been acquired through natural selection, but from their rarity they need not here be further considered. In all ordinary 316cases an inequality would be no advantage or disadvantage to certain individuals more than to others; and therefore it could hardly have resulted from natural selection. We must attribute the inequality to the direct action of those unknown conditions, which with mankind lead to the males being born in a somewhat larger excess in certain countries than in others, or which cause the proportion between the sexes to differ slightly in legitimate and illegitimate births.

Let us now take the case of a species producing from the unknown causes just alluded to, an excess of one sex—we will say of males—these being superfluous and useless, or nearly useless. Could the sexes be equalised through natural selection? We may feel sure, from all characters being variable, that certain pairs would produce a somewhat less excess of males over females than other pairs. The former, supposing the actual number of the offspring to remain constant, would necessarily produce more females, and would therefore be more productive. On the doctrine of chances a greater number of the offspring of the more productive pairs would survive; and these would inherit a tendency to procreate fewer males and more females. Thus a tendency towards the equalisation of the sexes would be brought about. But our supposed species would by this process be rendered, as just remarked, more productive; and this would in many cases be far from an advantage; for whenever the limit to the numbers which exist, depends, not on destruction by enemies, but on the amount of food, increased fertility will lead to severer competition and to most of the survivors being badly fed. In this case, if the sexes were equalised by an increase in the number of the females, a simultaneous decrease in the total number of the offspring would be beneficial, or even necessary, for the existence of the species; and317 this, I believe, could be effected through natural selection in the manner hereafter to be described. The same train of reasoning is applicable in the above, as well as in the following case, if we assume that females instead of males are produced in excess, for such females from not uniting with males would be superfluous and useless. So it would be with polygamous species, if we assume the excess of females to be inordinately great.

An excess of either sex, we will again say of the males, could, however, apparently be eliminated through natural selection in another and indirect manner, namely by an actual diminution of the males, without any increase of the females, and consequently without any increase in the productiveness of the species. From the variability of all characters, we may feel assured that some pairs, inhabiting any locality, would produce a rather smaller excess of superfluous males, but an equal number of productive females. When the offspring from the more and the less male-productive parents were all mingled together, none would have any direct advantage over the others; but those that produced few superfluous males would have one great indirect advantage, namely that their ova or embryos would probably be larger and finer, or their young better nurtured in the womb and afterwards. We see this principle illustrated with plants; as those which bear a vast number of seed produce small ones; whilst those which bear comparatively few seeds, often produce large ones well-stocked with nutriment for the use of the seedlings.408 Hence the offspring of the parents which 318had wasted least force in producing superfluous males would be the most likely to survive, and would inherit the same tendency not to produce superfluous males, whilst retaining their full fertility in the production of females. So it would be with the converse case of the female sex. Any slight excess, however, of either sex could hardly be checked in so indirect a manner. Nor indeed has a considerable inequality between the sexes been always prevented, as we have seen in some of the cases given in the previous discussion. In these cases the unknown causes which determine the sex of the embryo, and which under certain conditions lead to the production of one sex in excess over the other, have not been mastered by the survival of those varieties which were subjected to the least waste of organised matter and force by the production of superfluous individuals of either sex.

Nevertheless we may conclude that natural selection will always tend, though sometimes inefficiently, to equalise the relative numbers of the two sexes.

Having said this much on the equalisation of the sexes, it may be well to add a few remarks on the regulation through natural selection of the ordinary fertility of species. Mr. Herbert Spencer has shewn in an able discussion409 that with all organisms a ratio exists between what he calls individuation and genesis; whence it follows that beings which consume much matter or force in their growth, complicated structure or activity, or which produce ova and embryos of large size, or which expend much energy in nurturing their young, cannot be so productive as beings of an opposite nature. Mr. Spencer further shews that minor differences in fertility will be regulated through natural selection. Thus 319the fertility of each species will tend to increase, from the more fertile pairs producing a larger number of offspring, and these from their mere number will have the best chance of surviving, and will transmit their tendency to greater fertility. The only check to a continued augmentation of fertility in each organism seems to be either the expenditure of more power and the greater risks run by the parents that produce a more numerous progeny, or the contingency of very numerous eggs and young being produced of smaller size, or less vigorous, or subsequently not so well nurtured. To strike a balance in any case between the disadvantages which follow from the production of a numerous progeny, and the advantages (such as the escape of at least some individuals from various dangers) is quite beyond our power of judgment.

When an organism has once been rendered extremely fertile, how its fertility can be reduced through natural selection is not so clear as how this capacity was first acquired. Yet it is obvious that if individuals of a species, from a decrease of their natural enemies, were habitually reared in larger numbers than could be supported, all the members would suffer. Nevertheless the offspring from the less fertile parents would have no direct advantage over the offspring from the more fertile parents, when all were mingled together in the same district. All the individuals would mutually tend to starve each other. The offspring indeed of the less fertile parents would lie under one great disadvantage, for from the simple fact of being produced in smaller numbers, they would be the most liable to extermination. Indirectly, however, they would partake of one great advantage; for under the supposed condition of severe competition, when all were pressed for food, it is extremely probable that those individuals which from320 some variation in their constitution produced fewer eggs or young, would produce them of greater size or vigour; and the adults reared from such eggs or young would manifestly have the best chance of surviving, and would inherit a tendency towards lessened fertility. The parents, moreover, which had to nourish or provide for fewer offspring would themselves be exposed to a less severe strain in the struggle for existence, and would have a better chance of surviving. By these steps, and by no others as far as I can see, natural selection under the above conditions of severe competition for food, would lead to the formation of a new race less fertile, but better adapted for survival, than the parent-race.

CHAPTER IX.

Secondary Sexual Characters in the Lower Classes of the Animal Kingdom.

These characters absent in the lowest classes—Brilliant colours—Mollusca—Annelids—Crustacea, secondary sexual characters strongly developed; dimorphism; colour; characters not acquired before maturity—Spiders, sexual colours of; stridulation by the males—Myriapoda.

In the lowest classes the two sexes are not rarely united in the same individual, and therefore secondary sexual characters cannot be developed. In many cases in which the two sexes are separate, both are permanently attached to some support, and the one cannot search or struggle for the other. Moreover it is almost certain that these animals have too imperfect senses and much too low mental powers to feel mutual rivalry, or to appreciate each other's beauty or other attractions.

Hence in these classes, such as the Protozoa, Cœlenterata, Echinodermata, Scolecida, true secondary sexual characters do not occur; and this fact agrees with the belief that such characters in the higher classes have been acquired through sexual selection, which depends on the will, desires, and choice of either sex. Nevertheless some few apparent exceptions occur; thus, as I hear from Dr. Baird, the males of certain Entozoa, or internal parasitic worms, differ slightly in colour from the females; but we have no reason to suppose that such differences have been augmented through sexual selection.

322Many of the lower animals, whether hermaphrodites or with the sexes separate, are ornamented with the most brilliant tints, or are shaded and striped in an elegant manner. This is the case with many corals and sea-anemonies (Actineæ), with some jelly-fish (Medusæ, Porpita, &c.), with some Planariæ, Ascidians, numerous Star-fishes, Echini, &c.; but we may conclude from the reasons already indicated, namely the union of the two sexes in some of these animals, the permanently affixed condition of others, and the low mental powers of all, that such colours do not serve as a sexual attraction, and have not been acquired through sexual selection. With the higher animals the case is very different; for with them when one sex is much more brilliantly or conspicuously coloured than the other, and there is no difference in the habits of the two sexes which will account for this difference, we have reason to believe in the influence of sexual selection; and this belief is strongly confirmed when the more ornamented individuals, which are almost always the males, display their attractions before the other sex. We may also extend this conclusion to both sexes, when coloured alike, if their colours are plainly analogous to those of one sex alone in certain other species of the same group.

How, then, are we to account for the beautiful or even gorgeous colours of many animals in the lowest classes? It appears very doubtful whether such colours usually serve as a protection; but we are extremely liable to err in regard to characters of all kinds in relation to protection, as will be admitted by every one who has read Mr. Wallace's excellent essay on this subject. It would not, for instance, at first occur to any one that the perfect transparency of the Medusæ, or jelly-fishes, was of the highest service to them as a323protection; but when we are reminded by Häckel that not only the medusæ but many floating mollusca, crustaceans, and even small oceanic fishes partake of this same glass-like structure, we can hardly doubt that they thus escape the notice of pelagic birds and other enemies.

Notwithstanding our ignorance how far colour in many cases serves as a protection, the most probable view in regard to the splendid tints of many of the lowest animals seems to be that their colours are the direct result either of the chemical nature or the minute structure of their tissues, independently of any benefit thus derived. Hardly any colour is finer than that of arterial blood; but there is no reason to suppose that the colour of the blood is in itself any advantage; and though it adds to the beauty of the maiden's cheek, no one will pretend that it has been acquired for this purpose. So again with many animals, especially the lower ones, the bile is richly coloured; thus the extreme beauty of the Eolidæ (naked sea-slugs) is chiefly due, as I am informed by Mr. Hancock, to the biliary glands seen through the translucent integuments; this beauty being probably of no service to these animals. The tints of the decaying leaves in an American forest are described by every one as gorgeous; yet no one supposes that these tints are of the least advantage to the trees. Bearing in mind how many substances closely analogous to natural organic compounds have been recently formed by chemists, and which exhibit the most splendid colours, it would have been a strange fact if substances similarly coloured had not often originated, independently of any useful end being thus gained, in the complex laboratory of living organisms.

324The sub-kingdom of the Mollusca.—Throughout this great division (taken in its largest acceptation) of the animal kingdom, secondary sexual characters, such as we are here considering, never, as far as I can discover, occur. Nor could they be expected in the three lowest classes, namely in the Ascidians, Polyzoa, and Brachiopods (constituting the Molluscoida of Huxley), for most of these animals are permanently affixed to a support or have their sexes united in the same individual. In the Lamellibranchiata, or bivalve shells, hermaphroditism is not rare. In the next higher class of the Gasteropoda, or marine univalve shells, the sexes are either united or separate. But in this latter case the males never possess special organs for finding, securing, or charming the females, or for fighting with other males. The sole external difference between the sexes consists, as I am informed by Mr. Gwyn Jeffreys, in the shell sometimes differing a little in form; for instance, the shell of the male periwinkle (Littorina littorea) is narrower and has a more elongated spire than that of the female. But differences of this nature, it may be presumed, are directly connected with the act of reproduction or with the development of the ova.

The Gasteropoda, though capable of locomotion and furnished with imperfect eyes, do not appear to be endowed with sufficient mental powers for the members of the same sex to struggle together in rivalry, and thus to acquire secondary sexual characters. Nevertheless with the pulmoniferous gasteropods, or land-shells, the pairing is preceded by courtship; for these animals, though hermaphrodites, are compelled by their structure to pair together. Agassiz remarks,410 "Quiconque a eu l'occasion d'observer les amours des lima325çons, ne saurait mettre en doute la séduction déployée dans les mouvements et les allures qui préparent et accomplissent le double embrassement de ces hermaphrodites." These animals appear also susceptible of some degree of permanent attachment: an accurate observer, Mr. Lonsdale, informs me that he placed a pair of land-shells (Helix pomatia), one of which was weakly, into a small and ill-provided garden. After a short time the strong and healthy individual disappeared, and was traced by its track of slime over a wall into an adjoining well-stocked garden. Mr. Lonsdale concluded that it had deserted its sickly mate; but after an absence of twenty-four hours it returned, and apparently communicated the result of its successful exploration, for both then started along the same track and disappeared over the wall.

Even in the highest class of the Mollusca, namely the Cephalopoda or cuttle-fishes, in which the sexes are separate, secondary sexual characters of the kind which we are here considering, do not, as far as I can discover, occur. This is a surprising circumstance, as these animals possess highly-developed sense-organs and have considerable mental powers, as will be admitted by every one who has watched their artful endeavours to escape from an enemy.411 Certain Cephalopoda, however, are characterised by one extraordinary sexual character, namely, that the male element collects within one of the arms or tentacles, which is then cast off, and, clinging by its sucking-discs to the female, lives for a time an independent life. So completely does the cast-off arm resemble a separate animal, that it was described by Cuvier as a parasitic worm under the name 326of Hectocotyle. But this marvellous structure may be classed as a primary rather than as a secondary sexual character.

Although with the Mollusca sexual selection does not seem to have come into play; yet many univalve and bivalve shells, such as volutes, cones, scallops, &c., are beautifully coloured and shaped. The colours do not appear in most cases to be of any use as a protection; they are probably the direct result, as in the lowest classes, of the nature of the tissues; the patterns and the sculpture of the shell depending on its manner of growth. The amount of light seems to a certain extent to be influential; for although, as repeatedly stated by Mr. Gwyn Jeffreys, the shells of some species living at a profound depth are brightly coloured, yet we generally see the lower surfaces and the parts covered by the mantle less highly coloured than the upper and exposed surfaces.412 In some cases, as with shells living amongst corals or brightly-tinted sea-weeds, the bright colours may serve as a protection. But many of the nudibranch mollusca, or sea-slugs, are as beautifully coloured as any shells, as may be seen in Messrs. Alder and Hancock's magnificent work; and from information kindly given me by Mr. Hancock, it is extremely doubtful whether these colours usually serve as a protection. With some species this may be the case, as with one which lives on the green leaves of algæ, and is itself bright-green. But many brightly-coloured, white or otherwise conspicuous species, do not seek concealment; whilst again some equally conspicuous species, as well as other dull-coloured kinds, live under stones and in 327dark recesses. So that with these nudibranch molluscs, colour apparently does not stand in any close relation to the nature of the places which they inhabit.

These naked sea-slugs are hermaphrodites, yet they pair together, as do land-snails, many of which have extremely pretty shells. It is conceivable that two hermaphrodites, attracted by each others' greater beauty, might unite and leave offspring which would inherit their parents' greater beauty. But with such lowly-organised creatures this is extremely improbable. Nor is it at all obvious how the offspring from the more beautiful pairs of hermaphrodites would have any advantage, so as to increase in numbers, over the offspring of the less beautiful, unless indeed vigour and beauty generally coincided. We have not here a number of males becoming mature before the females, and the more beautiful ones selected by the more vigorous females. If, indeed, brilliant colours were beneficial to an hermaphrodite animal in relation to its general habits of life, the more brightly-tinted individuals would succeed best and would increase in number; but this would be a case of natural and not of sexual selection.

Sub-kingdom of the Vermes or Annulosa: Class, Annelida (or Sea-worms).—In this class, although the sexes (when separate) sometimes differ from each other in characters of such importance that they have been placed under distinct genera or even families, yet the differences do not seem of the kind which can be safely attributed to sexual selection. These animals, like those in the preceding classes, apparently stand too low in the scale, for the individuals of either sex to exert any choice in selecting a partner, or for the individuals of the same sex to struggle together in rivalry.

328Sub-kingdom of the Arthropoda: Class, Crustacea.—In this great class we first meet with undoubted secondary sexual characters, often developed in a remarkable manner. Unfortunately the habits of crustaceans are very imperfectly known, and we cannot explain the uses of many structures peculiar to one sex. With the lower parasitic species the males are of small size, and they alone are furnished with perfect swimming-legs, antennæ and sense-organs; the females being destitute of these organs, with their bodies often consisting of a mere distorted mass. But these

c. Ditto of female.differences between the two sexes are no doubt related to their widely different habits of life, and consequently do not concern us. In various crustaceans, belonging to distinct families, the anterior antennæ are furnished with peculiar thread-like bodies, which are believed to act as smelling-organs, and these are much more numerous in the males than in the females. As the males, without any unusual development of their olfactory organs, would almost certainly be able sooner or later to find the females, the increased number of the smelling-threads has probably been acquired through sexual selection, by the better provided males having been the most successful in finding partners and in leaving offspring. Fritz Müller has described a remarkable dimorphic species of Tanais, in which the male is represented by two distinct forms, never graduating into each other. In the one form the male is furnished with more numerous smelling-threads, and in the other form with more powerful and more elongated chelæ or pincers which serve to hold the female. Fritz Müller suggests that these differences between the two male forms of the same species must have originated in certain individuals having varied in the number of the smelling-threads, whilst other individuals varied in the shape and size of329 their chelæ; so that of the former, those which were best able to find the female, and of the latter, those which were best able to hold her when found, have left the greater number of progeny to inherit their respective advantages.413

In some of the lower crustaceans, the right-hand anterior antenna of the male differs greatly in structure from the left-hand one, the latter resembling in its simple tapering joints the antennæ of the female. In the male the modified antenna is either swollen in the middle or angularly bent, or converted (fig. 3) into an elegant, and sometimes wonderfully complex, prehensile organ.414 It serves, as I hear from Sir J. Lubbock, to hold the female, and for this same purpose one of the two posterior legs (b) on the same side of the body is converted into a forceps. In another family the inferior or posterior antennæ are "curiously zigzagged" in the males alone.

330

In the higher crustaceans the anterior legs form a pair of chelæ or pincers, and these are generally larger in the male than in the female. In many species the chelæ on the opposite sides of the body are of unequal size, the right-hand one being, as I am informed by Mr. C. Spence Bate, generally, though not invariably, the largest. This inequality is often much greater in the male than in the female. The two chelæ also often differ in structure (figs. 4 and 5), the smaller one resembling those of the female. What advantage331 is gained by their inequality in size on the opposite sides of the body, and by the inequality being much greater in the male than in the female; and why, when they are of equal size, both are often much larger in the male than in the female, is not known. The chelæ are sometimes of such length and size that they cannot possibly be used, as I hear from Mr. Spence Bate, for carrying food to the mouth. In the males of certain freshwater prawns (Palæmon) the right leg is actually longer than the whole body.415 It is probable that the great size of one leg with its chelæ may aid the male in fighting with his rivals; but this use will not account for their inequality in the female on the opposite sides of the body. In Gelasimus, according to a statement quoted by Milne-Edwards,416 the male and female live in the same burrow, which is worth notice, as shewing that they pair, and the male closes the mouth of the burrow with one of its chelæ, which is enormously developed; so that here it indirectly serves as a means of defence. Their main use, however, probably is to seize and to secure the female, and this in some instances, as with Gammarus, is known to be the case. The sexes, however, of the common shore-crab (Carcinus mænas), as Mr. Spence Bate informs me, unite directly after the female has moulted her hard shell, and when she is so soft that she would be injured if seized by the strong pincers of the male; but as she is caught and carried about by the male previously to the act of moulting, she could then be seized with impunity.

Fritz Müller states that certain species of Melita are 332distinguished from all other amphipods by the females having "the coxal lamellæ of the penultimate pair of feet produced into hook-like processes, of which the males lay hold with the hands of the first pair." The development of these hook-like processes probably resulted from those females which were the most securely held during the act of reproduction, having left the largest number of offspring. Another Brazilian amphipod (Orchestia Darwinii, fig. 7) is described by Fritz Müller, as presenting a case of dimorphism, like that of Tanais; for there are two male forms, which differ in the structure of their chelæ.417 As chelæ of either shape would certainly have sufficed to hold the female, for both are now used for this purpose, the two male forms probably originated, by some having varied in one manner and some in another; both forms having derived certain special, but nearly equal advantages, from their differently shaped organs.

It is not known that male crustaceans fight together for the possession of the females, but this is probable; for with most animals when the male is larger than the female, he seems to have acquired his greater size by having conquered during many generations other males. Now Mr. Spence Bate informs me that in most of the crustacean orders, especially in the highest or the Brachyura, the male is larger than the female; the parasitic genera, however, in which the sexes follow different habits of life, and most of the Entomostraca must be excepted. The chelæ of many crustaceans are weapons well adapted for fighting. Thus a Devil-crab (Portunus puber) was seen by a son of Mr. Bate fighting with a Carcinus mænas, and the latter was soon thrown on its back, and had every limb torn from its body. 333When several males of a Brazilian Gelasimus, a species furnished with immense pincers, were placed together by Fritz Müller in a glass vessel, they mutilated and killed each other. Mr. Bate put a large male Carcinus mænas into a pan of water, inhabited by a female paired with a smaller male; the latter was soon dispossessed, but, as Mr. Bate adds, "if they fought, the victory334 was a bloodless one, for I saw no wounds." This same naturalist separated a male sand-skipper (so common on our sea-shores), Gammarus marinus, from its female, both of which were imprisoned in the same vessel with many individuals of the same species. The female being thus divorced joined her comrades. After an interval the male was again put into the same vessel and he then, after swimming about for a time, dashed into the crowd, and without any fighting at once took away his wife. This fact shews that in the Amphipoda, an order low in the scale, the males and females recognise each other, and are mutually attached.

The mental powers of the Crustacea are probably higher than might have been expected. Any one who has tried to catch one of the shore-crabs, so numerous on many tropical coasts, will have perceived how wary and alert they are. There is a large crab (Birgos latro), found on coral islands, which makes at the bottom of a deep burrow a thick bed of the picked fibres of the cocoa-nut. It feeds on the fallen fruit of this tree by tearing off the husk, fibre by fibre; and it always begins at that end where the three eye-like depressions are situated. It then breaks through one of these eyes by hammering with its heavy front pincers, and turning round, extracts the albuminous core with its narrow posterior pincers. But these actions are probably instinctive, so that they would be performed as well by a young as by an old animal. The following case, however, can hardly be so considered: a trustworthy naturalist, Mr. Gardner,418 whilst watching a shore-crab (Gelasimus) making its burrow, 335threw some shells towards the hole. One rolled in, and three other shells remained within a few inches of the mouth. In about five minutes the crab brought out the shell which had fallen in, and carried it away to the distance of a foot; it then saw the three other shells lying near, and evidently thinking that they might likewise roll in, carried them to the spot where it had laid the first. It would, I think, be difficult to distinguish this act from one performed by man by the aid of reason.

With respect to colour which so often differs in the two sexes of animals belonging to the higher classes, Mr. Spence Bate does not know of any well-marked instances with our British crustaceans. In some cases, however, the male and female differ slightly in tint, but Mr. Bate thinks not more than may be accounted for by their different habits of life, such as by the male wandering more about and being thus more exposed to the light. In a curious Bornean crab, which inhabits sponges, Mr. Bate could always distinguish the sexes by the male not having the epidermis so much rubbed off. Dr. Power tried to distinguish by colour the sexes of the species which inhabit the Mauritius, but always failed, except with one species of Squilla, probably the S. stylifera, the male of which is described as being "of a beautiful blueish-green," with some of the appendages cherry-red, whilst the female is clouded with brown and grey, "with the red about her much less vivid than in the male."419 In this case, we may suspect the agency of sexual selection. With Saphirina (an oceanic genus of Entomostraca, and therefore low in the scale) the males are furnished with 336minute shields or cell-like bodies, which exhibit beautiful changing colours; these being absent in the females, and in the case of one species in both sexes.420 It would, however, be extremely rash to conclude that these curious organs serve merely to attract the females. In the female of a Brazilian species of Gelasimus, the whole body, as I am informed by Fritz Müller, is of a nearly uniform greyish-brown. In the male the posterior part of the cephalo-thorax is pure white, with the anterior part of a rich green, shading into dark brown; and it is remarkable that these colours are liable to change in the course of a few minutes—the white becoming dirty grey or even black, the green "losing much of its brilliancy." The males apparently are much more numerous than the females. It deserves especial notice that they do not acquire their bright colours until they become mature. They differ also from the females in the larger size of their chelæ. In some species of the genus, probably in all, the sexes pair and inhabit the same burrow. They are also, as we have seen, highly intelligent animals. From these various considerations it seems highly probable that the male in this species has become gaily ornamented in order to attract or excite the female.

It has just been stated that the male Gelasimus does not acquire his conspicuous colours until mature and nearly ready to breed. This seems the general rule in the whole class with the many remarkable differences in structure between the two sexes. We shall hereafter find the same law prevailing throughout the great sub-kingdom of the Vertebrata, and in all cases it is eminently distinctive of characters which have been 337acquired through sexual selection. Fritz Müller421 gives some striking instances of this law; thus the male sand-hopper (Orchestia) does not acquire his large claspers, which are very differently constructed from those of the female, until nearly full-grown; whilst young his claspers resemble those of the female. Thus, again, the male Brachyscelus possesses, like all other amphipods, a pair of posterior antennæ; the female, and this is a most extraordinary circumstance, is destitute of them, and so is the male as long as he remains immature.

Class, Arachnida (Spiders).—The males are often darker, but sometimes lighter than the females, as may be seen in Mr. Blackwall's magnificent work.422 In some species the sexes differ conspicuously from each other in colour; thus the female of Sparassus smaragdulus is dullish-green; whilst the adult male has the abdomen of a fine yellow, with three longitudinal stripes of rich red. In some species of Thomisus the two sexes closely resemble each other; in others they differ much; thus in T. citreus the legs and body of the female are pale-yellow or green, whilst the front legs of the male are reddish-brown: in T. floricolens, the legs of the female are pale-green, those of the male being ringed in a conspicuous manner with various tints. Numerous analogous cases could be given in the genera Epeira, Nephila, Philodromus, Theridion, Linyphia, &c. It is often difficult to say which of the two sexes departs most from the ordinary coloration of the genus to which the species belong; but Mr. Blackwall 338thinks that, as a general rule, it is the male. Both sexes whilst young, as I am informed by the same author, usually resemble each other; and both often undergo great changes in colour during their successive moults before arriving at maturity. In other cases the male alone appears to change colour. Thus the male of the above-mentioned brightly-coloured Sparassus at first resembles the female and acquires his peculiar tints only when nearly adult. Spiders are possessed of acute senses, and exhibit much intelligence. The females often shew, as is well known, the strongest affection for their eggs, which they carry about enveloped in a silken web. On the whole it appears probable that well-marked differences in colour between the sexes have generally resulted from sexual selection, either on the male or female side. But doubts may be entertained on this head from the extreme variability in colour of some species, for instance of Theridion lineatum, the sexes of which differ when adult; this great variability indicates that their colours have not been subjected to any form of selection.

Mr. Blackwall does not remember to have seen the males of any species fighting together for the possession of the female. Nor, judging from analogy, is this probable; for the males are generally much smaller than the females, sometimes to an extraordinary degree.423 Had the males been in the habit of fighting together, they would, it is probable, have gradually339acquired greater size and strength. Mr. Blackwall has sometimes seen two or more males on the same web with a single female; but their courtship is too tedious and prolonged an affair to be easily observed. The male is extremely cautious in making his advances, as the female carries her coyness to a dangerous pitch. De Geer saw a male that "in the midst of his preparatory caresses was seized by the object of his attractions, enveloped by her in a web and then devoured, a sight which, as he adds, filled him with horror and indignation."424

Westring has made the interesting discovery that the males of several species of Theridion425 have the power of making a stridulating sound (like that made by many beetles and other insects, but feebler), whilst the females are quite mute. The apparatus consists of a serrated ridge at the base of the abdomen, against which the hard hinder part of the thorax is rubbed; and of this structure not a trace could be detected in the females. From the analogy of the Orthoptera and Homoptera, to be described in the next chapter, we may feel almost sure that the stridulation serves, as Westring remarks, either to call or to excite the female; and this is the first case in the ascending scale of the animal kingdom, known to me, of sounds emitted for this purpose.

Class, Myriapoda.—In neither of the two orders in this class, including the millipedes and centipedes, 340can I find any well-marked instances of sexual differences such as more particularly concern us. In Glomeris limbata, however, and perhaps in some few other species, the males differ slightly in colour from the females; but this Glomeris is a highly variable species. In the males of the Diplopoda, the legs belonging to one of the anterior segments of the body, or to the posterior segment, are modified into prehensile hooks which serve to secure the female. In some species of Iulus the tarsi of the male are furnished with membranous suckers for the same purpose. It is a much more unusual circumstance, as we shall see when we treat of Insects, that it is the female in Lithobius which is furnished with prehensile appendages at the extremity of the body for holding the male.426

CHAPTER X.
Secondary Sexual Characters of Insects.

Diversified structures possessed by the males for seizing the females—Differences between the sexes, of which the meaning is not understood—Difference in size between the sexes—Thysanura—Diptera—Hemiptera—Homoptera, musical powers possessed by the males alone—Orthoptera, musical instruments of the males, much diversified in structure; pugnacity; colours—Neuroptera, sexual differences in colour—Hymenoptera, pugnacity and colours—Coleoptera, colours; furnished with great horns, apparently as an ornament; battles; stridulating organs generally common to both sexes.

In the immense class of insects the sexes sometimes differ in their organs for locomotion, and often in their sense-organs, as in the pectinated and beautifully plumose antennæ of the males of many species. In one of the Ephemeræ, namely Chloëon, the male has great pillared eyes, of which the female is entirely destitute.427 The ocelli are absent in the females of certain other insects, as in the Mutillidæ, which are likewise destitute of wings. But we are chiefly concerned with structures by which one male is enabled to conquer another, either in battle or courtship, through his strength, pugnacity, ornaments, or music. The innumerable contrivances, therefore, by which the male is able to seize the female, may be briefly passed over. Besides the complex structures at the apex of the abdomen, which ought perhaps to be ranked as primary 342organs,428 "it is astonishing," as Mr. B. D. Walsh429 has remarked, "how many different organs are worked in by nature, for the seemingly insignificant object of enabling the male to grasp the female firmly." The mandibles or jaws are sometimes used for this purpose; thus the male Corydalis cornutus (a neuropterous insect in some degree allied to the Dragon-flies, &c.) has immense curved jaws, many times longer than those of the female; and they are smooth instead of being toothed, by which means he is enabled to seize her without injury.430 One of the stag-beetles of North America (Lucanus elaphus) uses his jaws, which are much larger than those of the female, for the same purpose, but probably likewise for fighting. In one of the sand-wasps (Ammophila) the jaws in the two sexes are closely alike, but are used for widely different purposes; the males, as Professor Westwood observes, "are exceedingly ardent, seizing their partners round the neck with their sickle-shaped jaws;"431 whilst the females use 343these organs for burrowing in sand-banks and making their nests.

The tarsi of the front-legs are dilated in many male beetles, or are furnished with broad cushions of hairs; and in many genera of water-beetles they are armed with a round flat sucker, so that the male may adhere to the slippery body of the female. It is a much more unusual circumstance that the females of some water-beetles Fig. 8. Crabro cribrarius. Upper figure, male: lower figure, female.(Dytiscus) have their elytra deeply grooved, and in Acilius sulcatus thickly set with hairs, as an aid to the male. The females of some other water-beetles (Hydroporus) have their elytra punctured for the same object.432 In the male of Crabro cribrarius (fig. 8.), it is the tibia which is dilated into a broad horny plate, with minute membraneous dots, giving to it a singular appearance like that of a riddle.433 In the male of Penthe (a genus of beetles) a few of the middle joints of the antennæ are dilated and furnished on the inferior surface 344with cushions of hair, exactly like those on the tarsi of the Carabidæ, "and obviously for the same end." In male dragon-flys, "the appendages at the tip of the tail are modified in an almost infinite variety of curious patterns to enable them to embrace the neck of the female." Lastly in the males of many insects, the legs are furnished with peculiar spines, knobs or spurs; or the whole leg is bowed or thickened, but this is by no means invariably a sexual character; Fig. 9. Taphroderes distortus (much enlarged). Upper figure, male; lower figure, female.or one pair, or all three pairs are elongated, sometimes to an extravagant length.434

In all the orders, the sexes of many species present differences, of which the meaning is not understood. One curious case is that of a beetle (fig. 9), the male of which has the left mandible much enlarged; so that the mouth is greatly distorted. In another Carabidous beetle, the Eurygnathus,435 we have the unique case, as far as known to Mr. Wollaston, of the head of the female being much broader and larger, though in a variable degree, than that of the male. Any number of such cases could be given. They abound in the Lepidoptera: one of the most extraordinary is that certain male butterflies have their fore-legs more or 345less atrophied, with the tibiæ and tarsi reduced to mere rudimentary knobs. The wings, also, in the two sexes often differ in neuration,436 and sometimes considerably in outline, as in the Aricoris epitus, which was shown to me in the British Museum by Mr. A. Butler. The males of certain South American butterflies have tufts of hair on the margins of the wings, and horny excrescences on the discs of the posterior pair.437 In several British butterflies, the males alone, as shewn by Mr. Wonfor, are in parts clothed with peculiar scales.

The purpose of the luminosity in the female glow-worm is likewise not understood; for it is very doubtful whether the primary use of the light is to guide the male to the female. It is no serious objection to this latter belief that the males emit a feeble light; for secondary sexual characters proper to one sex are often developed in a slight degree in the other sex. It is a more valid objection that the larvæ shine, and in some species brilliantly: Fritz Müller informs me that the most luminous insect which he ever beheld in Brazil, was the larva of some beetle. Both sexes of certain luminous species of Elater emit light. Kirby and Spence suspect that the phosphorescence serves to frighten and drive away enemies.

Difference in Size between the Sexes.—With insects of all kinds the males are commonly smaller than the females;438 and this difference can often be detected even in the larval state. So considerable is the difference 346between the male and female cocoons of the silk-moth (Bombyx mori), that in France they are separated by a particular mode of weighing.439 In the lower classes of the animal kingdom, the greater size of the females seems generally to depend on their developing an enormous number of ova; and this may to a certain extent hold good with insects. But Dr. Wallace has suggested a much more probable explanation. He finds, after carefully attending to the development of the caterpillars of Bombyx cynthia and Yamamai, and especially of some dwarfed caterpillars reared from a second brood on unnatural food, "that in proportion as the individual moth is finer, so is the time required for its metamorphosis longer; and for this reason the female, which is the larger and heavier insect, from having to carry her numerous eggs, will be preceded by the male, which is smaller and has less to mature."440 Now as most insects are short-lived, and as they are exposed to many dangers, it would manifestly be advantageous to the female to be impregnated as soon as possible. This end would be gained by the males being first matured in large numbers ready for the advent of the females; and this again would naturally follow, as Mr. A. E. Wallace has remarked,441 through natural selection; for the smaller males would be first matured, and thus would procreate a large number of offspring which would inherit the reduced size of their male parents, whilst the larger males from being matured later would leave fewer offspring.

There are, however, exceptions to the rule of male insects being smaller than the females; and some of 347these exceptions are intelligible. Size and strength would be an advantage to the males, which fight for the possession of the female; and in these cases the males, as with the stag-beetle (Lucanus), are larger than the females. There are, however, other beetles which are not known to fight together, of which the males exceed the females in size; and the meaning of this fact is not known; but in some of these cases, as with the huge Dynastes and Megasoma, we can at least see that there would be no necessity for the males to be smaller than the females, in order to be matured before them, for these beetles are not short-lived, and there would be ample time for the pairing of the sexes. So, again, male dragon-flies (Libellulidæ) are sometimes sensibly larger, and never smaller, than the females;442 and they do not, as Mr. MacLachlan believes, generally pair with the females, until a week or fortnight has elapsed, and until they have assumed their proper masculine colours. But the most curious case, shewing on what complex and easily-overlooked relations, so trifling a character as a difference in size between the sexes may depend, is that of the aculeate Hymenoptera; for Mr. F. Smith informs me that throughout nearly the whole of this large group the males, in accordance with the general rule, are smaller than the females and emerge about a week before them; but amongst the Bees, the males of Apis mellifica, Anthidium manicatum and Anthophora acervorum, and amongst the Fossores, the males of the Methoca ichneumonides, are larger than the females. The explanation of this anomaly is that a marriage-flight is absolutely necessary 348with these species, and the males require great strength and size in order to carry the females through the air. Increased size has here been acquired in opposition to the usual relation between size and the period of development, for the males, though larger, emerge before the smaller females.

We will now review the several Orders, selecting such facts as more particularly concern us. The Lepidoptera (Butterflies and Moths) will be retained for a separate chapter.

Order, Thysanura.—The members of this Order are lowly organised for their class. They are wingless, dull-coloured, minute insects, with ugly, almost misshapen heads and bodies. The sexes do not differ; but they offer one interesting fact, by showing that the males pay sedulous court to their females even low down in the animal scale. Sir J. Lubbock443 in describing the Smynthurus luteus, says: "it is very amusing to see these little creatures coquetting together. The male, which is much smaller than the female, runs round her, and they butt one another, standing face to face, and moving backward and forward like two playful lambs. Then the female pretends to run away and the male runs after her with a queer appearance of anger, gets in front and stands facing her again; then she turns coyly round, but he, quicker and more active, scuttles round too, and seems to whip her with his antennæ; then for a bit they stand face to face, play with their antennæ, and seem to be all in all to one another."

Order, Diptera (Flies).—The sexes differ little in colour. The greatest difference, known to Mr. F. Walker, 349is in the genus Bibio, in which the males are blackish or quite black, and the females obscure brownish-orange. The genus Elaphomyia, discovered by Mr. Wallace444 in New Guinea, is highly remarkable, as the males are furnished with horns, of which the females are quite destitute. The horns spring from beneath the eyes, and curiously resemble those of stags, being either branched or palmated. They equal in length the whole of the body in one of the species. They might be thought to serve for fighting, but as in one species they are of a beautiful pink colour, edged with black, with a pale central stripe, and as these insects have altogether a very elegant appearance, it is perhaps more probable that the horns serve as ornaments. That the males of some Diptera fight together is certain; for Prof. Westwood445 has several times seen this with some species of Tipula or Harry-long-legs. Many observers believe that when gnats (Culicidæ) dance in the air in a body, alternately rising and falling, the males are courting the females. The mental faculties of the Diptera are probably fairly well developed, for their nervous system is more highly developed than in most other Orders of insects.446

Order, Hemiptera (Field-Bugs).—Mr. J. W. Douglas, who has particularly attended to the British species, has kindly given me an account of their sexual differences. The males of some species are furnished with wings, whilst the females are wingless; the sexes differ in the form of the body and elytra; in the second joints of their antennæ and in their tarsi; but as the signification 350of these differences is quite unknown, they may be here passed over. The females are generally larger and more robust than the males. With British, and, as far as Mr. Douglas knows, with exotic species, the sexes do not commonly differ much in colour; but in about six British species the male is considerably darker than the female, and in about four other species the female is darker than the male. Both sexes of some species are beautifully marked with vermilion and black. It is doubtful whether these colours serve as a protection. If in any species the males had differed from the females in an analogous manner, we might have been justified in attributing such conspicuous colours to sexual selection with transference to both sexes.

Some species of Reduvidæ make a stridulating noise; and, in the case of Pirates stridulus, this is said447 to be effected by the movement of the neck within the pro-thoracic cavity. According to Westring, Reduvius personatus also stridulates. But I have not been able to learn any particulars about these insects; nor have I any reason to suppose that they differ sexually in this respect.

Order, Homoptera.—Every one who has wandered in a tropical forest must have been astonished at the din made by the male Cicadæ. The females are mute; as the Grecian poet Xenarchus says, "Happy the Cicadas live, since they all have voiceless wives." The noise thus made could be plainly heard on board the "Beagle," when anchored at a quarter of a mile from the shore of Brazil; and Captain Hancock says it can be heard at the distance of a mile. The Greeks formerly kept, and the Chinese now keep, these insects 351in cages for the sake of their song, so that it must be pleasing to the ears of some men.448 The Cicadidæ usually sing during the day; whilst the Fulgoridæ appear to be night-songsters. The sound, according to Landois,449 who has recently studied the subject, is produced by the vibration of the lips of the spiracles, which are set into motion by a current of air emitted from the tracheæ. It is increased by a wonderfully complex resounding apparatus, consisting of two cavities covered by scales. Hence the sound may truly be called a voice. In the female the musical apparatus is present, but very much less developed than in the male, and is never used for producing sound.

With respect to the object of the music, Dr. Hartman in speaking of the Cicada septemdecim of the United States, says,450 "the drums are now (June 6th and 7th, 1851) heard in all directions. This I believe to be the marital summons from the males. Standing in thick chestnut sprouts about as high as my head, where hundreds were around me, I observed the females coming around the drumming males." He adds, "this season (Aug. 1868) a dwarf pear-tree in my garden produced about fifty larvæ of Cic. pruinosa; and I several times noticed the females to alight near a male while he was uttering his clanging notes." Fritz Müller writes to me from S. Brazil that he has often listened to a musical contest between two or three males of a Cicada, having a particularly loud voice, and seated at a considerable distance from each other. As 352soon as the first had finished his song, a second immediately began; and after he had concluded, another began, and so on. As there is so much rivalry between the males, it is probable that the females not only discover them by the sounds emitted, but that, like female birds, they are excited or allured by the male with the most attractive voice.

I have not found any well-marked cases of ornamental differences between the sexes of the Homoptera. Mr. Douglas informs me that there are three British species, in which the male is black or marked with black bands, whilst the females are pale-coloured or obscure.

Order, Orthoptera.—The males in the three saltatorial families belonging to this Order are remarkable for their musical powers, namely the Achetidæ or crickets, the Locustidæ for which there is no exact equivalent name in English, and the Acridiidæ or grasshoppers. The stridulation produced by some of the Locustidæ is so loud that it can be heard during the night at the distance of a mile;[451] and that made by certain species is not unmusical even to the human ear, so that the Indians on the Amazons keep them in wicker cages. All observers agree that the sounds serve either to call or excite the mute females. But it has been noticed[452] that the male migratory locust of Russia (one of the Acridiidæ) whilst coupled with the female, stridulates from anger or jealousy when approached by another male. The house-cricket when surprised at night uses its voice to warn its fellows.[453] In North America the Katy-did (Platyphyllum concavum, 353one of the Locustidæ) is described[454] as mounting on the upper branches of a tree, and in the evening beginning "his noisy babble, while rival notes issue from the neighbouring trees, and the groves resound with the call of Katy-did-she-did, the live-long night." Mr. Bates, in speaking of the European field-cricket (one of the Achetidæ), says, "the male has been observed to place itself in the evening at the entrance of its burrow, and stridulate until a female approaches, when the louder notes are succeeded by a more subdued tone, whilst the successful musician caresses with   his   antennæ.

Right-hand figure, under side of part of the wing-nervure, much magnified, showing the teeth, st.

Left-hand figure, upper surface of wing-cover, with the projecting, smooth nervure,r., across which the teeth (st) are scraped.the mate he has won."[455] Dr. Scudder was able to excite one of these insects to answer him, by rubbing on a file with a quill.[456] In both sexes a remarkable auditory apparatus has been discovered by Von Siebold, situated in the front legs.[457]

In the three Families the sounds are differently produced. In the males Of the Achetidæ both wing-covers have the same structure; and this in the field-cricket (Gryllus campestris, fig. 10) consists, as de354scribed by Landois,[458] of from 131 to 138 sharp, transverse ridges or teeth (st) on the under side of one of the nervures of the wing-cover. This toothed nervure is rapidly scraped across a projecting, smooth, hard nervure (r) on the upper surface of the opposite wing.

Teeth of Nervure of Gryllus domesticus (from Landois).one wing is rubbed over the other, and then the movement is reversed. Both wings are raised a little at the same time, so as to increase the resonance. In some species the wing-covers of the males are furnished at the base with a talc-like plate.[459] I have here given a drawing (fig. 11) of the teeth on the under side of the nervure of another species of Gryllus, viz. G. domesticus.

In the Locustidæ the opposite wing-covers differ in structure (fig. 12), and cannot, as in the last family, be indifferently used in a reversed manner. The left wing, which acts as the bow of the fiddle, lies over the right wing which serves as the fiddle itself. One of the nervures (a) on the under surface of the former is finely serrated, and is scraped across the prominent nervures on the upper surface of the opposite or right wing. In our British Phasgonura viridissima it appeared to me that the serrated nervure is rubbed against the rounded hind corner of the opposite wing, the edge of which is thickened, coloured brown, and very sharp. In the right wing, but not in the left, there is a little plate, as transparent as talc, surrounded by nervures, and called the speculum. In Ephippiger vitium, a member of this same family, we have a curious 355subordinate modification; for the wing-covers are greatly reduced in size, but "the posterior part of the pro-thorax is elevated into a kind of dome over the wing-covers, and which has probably the effect of increasing the sound."460

We thus see that the musical apparatus is more differentiated or specialised in the Locustidæ, which includes I believe the most powerful performers in the Order, than in the Achetidæ, in which both wing-covers have the same structure and the same function.461 Landois, however, detected in one of the Locustidæ, namely in Decticus, a short and narrow row of small 356teeth, mere rudiments, on the inferior surface of the right wing-cover, which underlies the other and is never used as the bow. I observed the same rudimentary structure on the under side of the right wing-cover in Phasgonura viridissima. Hence we may with confidence infer that the Locustidæ are descended from a form, in which, as in the existing Achetidæ, both wing-covers had serrated nervures on the under surface, and could be indifferently used as the bow; but that in the Locustidæ the two wing-covers gradually became differentiated and perfected, on the principle of the division of labour, the one to act exclusively as the bow and the other as the fiddle. By what steps the more simple apparatus in the Achetidæ originated, we do not know, but it is probable that the basal portions of the wing-covers overlapped each other formerly as at present, and that the friction of the nervures produced a grating sound, as I find is now the case with the wing-covers of the females.462 A grating sound thus occasionally and accidentally made by the males, if it served them ever so little as a love-call to the females, might readily have been intensified through sexual selection by fitting variations in the roughness of the nervures having been continually preserved.

In the last and third Family, namely the Acridiidæ or grasshoppers, the stridulation is produced in a very different manner, and is not so shrill, according to Dr. Scudder, as in the preceding Families. The inner surface of the femur (fig. 13, r) is furnished with a longitudinal row of minute, elegant, lancet-shaped, elastic teeth, from 85 to 93 in number;463 and these are scraped 357across the sharp, projecting nervures on the wing-covers, which, are thus made to vibrate and resound. Harris464

Fig. 13, Hind-leg of Stenobothrus pratorum: r, the stridulating ridge; lower figure, the teeth, forming the ridge, much magnified (from Landois).says that when one of the males begins to play, he first "bends the shank of the hind-leg beneath, the thigh, where it is lodged in a furrow designed to receive it, and then draws the leg briskly up and down. He does not play both fiddles together, but alternately first upon one and then on the other." In many species, the base of the abdomen is hollowed out into a great cavity which is believed to act as a resounding board. In Pneumora (fig. 14), a S. African genus belonging to this same family, we meet with a new and remarkable modification: in the males a small notched ridge projects obliquely from each side of the abdomen, against which the hind femora are rubbed.465 As the male is furnished with wings, the female being wingless, it is remarkable that the thighs are not rubbed in the usual manner against the wing-covers; but this may perhaps be accounted for by the unusually small size of the hind-legs. I have not been able to examine the inner surface of the thighs, which, judging from analogy, would be finely serrated. The species of Pneumora have been more profoundly modified for the sake of stridulation than any other orthopterous insect; for 358in the male the whole body has been converted into a musical instrument, being distended with air, like a great pellucid bladder, so as to increase the resonance. Mr. Trimen informs me that at the Cape of Good Hope these insects make a wonderful noise during the night There is one exception to the rule that the females in these three Families are destitute of an efficient musical apparatus; for both sexes of Ephippiger (Locustidæ) are said466 to be thus provided. This case may 359be compared with that of the reindeer, in which species alone both sexes possess horns. Although the female orthoptera are thus almost invariably mute, yet Landois467 found rudiments of the stridulating organs on the femora of the female Acridiidæ, and similar rudiments on the under surface of the wing-covers of the female Achetidæ; but he failed to find any rudiments in the females of Decticus, one of the Locustidæ. In the Homoptera the mute females of Cicada, have the proper musical apparatus in an undeveloped state; and we shall hereafter meet in other divisions of the animal kingdom with innumerable instances of structures proper to the male being present in a rudimentary condition in the female. Such cases appear at first sight to indicate that both sexes were primordially constructed in the same manner, but that certain organs were subsequently lost by the females. It is, however, a more probable view, as previously explained, that the organs in question were acquired by the males and partially transferred to the females.

Landois has observed another interesting fact, namely that in the females of the Acridiidæ, the stridulating teeth on the femora remain throughout life in the same condition in which they first appear in both sexes during the larval state. In the males, on the other hand, they become fully developed and acquire their perfect structure at the last moult, when the insect is mature and ready to breed.

From the facts now given, we see that the means by which the males produce their sounds are extremely diversified in the Orthoptera, and are altogether different from those employed by the Homoptera. But throughout the animal kingdom we incessantly find the 360same object gained by the most diversified means; this being due to the whole organisation undergoing in the course of ages multifarious changes; and as part after part varies, different variations are taken advantage of for the same general purpose. The diversification of the means for producing sound in the three families of the Orthoptera and in the Homoptera, impresses the mind with the high importance of these structures to the males, for the sake of calling or alluring the females. We need feel no surprise at the amount of modification which the Orthoptera have undergone in this respect, as we now know, from Dr. Scudder's remarkable discovery,468 that there has been more than ample time. This naturalist has lately found a fossil insect in the Devonian formation of New Brunswick, which is furnished with "the well-known tympanum or stridulating apparatus of the male Locustidæ." This insect, though in most respects related to the Neuroptera, appears to connect, as is so often the case with very ancient forms, the two Orders of the Neuroptera and Orthoptera which are now generally ranked as quite distinct.

I have but little more to say on the Orthoptera. Some of the species are very pugnacious: when two male field-crickets (Gryllus campestris) are confined together, they fight till one kills the other; and the species of Mantis are described as manœuvring with their sword-like front-limbs, like hussars with their sabres. The Chinese keep these insects in little bamboo cages and match them like game-cocks.469 With respect to colour, some exotic locusts are beautifully ornamented; the posterior wings being marked with red, 361blue, and black; but as throughout the Order the two sexes rarely differ much in colour, it is doubtful whether they owe these bright tints to sexual selection. Conspicuous colours may be of use to these insects as a protection, on the principle to be explained in the next chapter, by giving notice to their enemies that they are unpalatable. Thus it has been observed470 that an Indian brightly-coloured locust was invariably rejected when offered to birds and lizards. Some cases, however, of sexual differences in colour in this Order are known. The male of an American cricket471 is described as being as white as ivory, whilst the female varies from almost white to greenish-yellow or dusky. Mr. Walsh informs me that the adult male of Spectrum femoratum (one of the Phasmidæ) "is of a shining brownish-yellow colour; the adult female being of a dull, opaque, cinereous-brown; the young of both sexes being green." Lastly, I may mention that the male of one curious kind of cricket472 is furnished with "a long membranous appendage, which falls over the face like a veil;" but whether this serves as an ornament is not known.

Order, Neuroptera.—Little need here be said, except in regard to colour. In the Ephemeridæ the sexes often differ slightly in their obscure tints;473 but it is not probable that the males are thus rendered attractive to the females. The Libellulidæ or dragon-flies are ornamented with splendid green, blue, yellow, and 362vermilion metallic tints; and the sexes often differ. Thus, the males of some of the Agrionidæ, as Prof. Westwood remarks474 "are of a rich blue with black wings, whilst the females are fine green with colourless wings." But inAgrion Ramburii these colours are exactly reversed in the two sexes.475 In the extensive N. American genus of Hetærina, the males alone have a beautiful carmine spot at the base of each wing. In Anax junius the basal part of the abdomen in the male is a vivid ultra-marine blue, and in the female grass-green. In the allied genus Gomphus, on the other hand, and in some other genera, the sexes differ but little in colour. Throughout the animal kingdom, similar cases of the sexes of closely-allied forms either differing greatly, or very little, or not at all, are of frequent occurrence. Although with many Libellulidæ there is so wide a difference in colour between the sexes, it is often difficult to say which is the most brilliant; and the ordinary coloration of the two sexes is exactly reversed, as we have just seen, in one species of Agrion. It is not probable that their colours in any case have been gained as a protection. As Mr. MacLachlan, who has closely attended to this family, writes to me, dragon-flies—the tyrants of the insect-world—are the least liable of any insect to be attacked by birds or other enemies. He believes that their bright colours serve as a sexual attraction. It deserves notice, as bearing on this subject, that certain dragon-flies appear to be attracted by particular colours: Mr. Patterson observed476 that the species of Agrionidæ, of which the males are blue, settled in numbers on the blue float of a fishing363line; whilst two other species were attracted by shining white colours.

It is an interesting fact, first observed by Schelver, that the males, in several genera belonging to two sub-families, when they first emerge from the pupal state are coloured exactly like the females; but that their bodies in a short time assume a conspicuous milky-blue tint, owing to the exudation of a kind of oil, soluble in ether and alcohol. Mr. MacLachlan believes that in the male of Libellula depressa this change of colour does not occur until nearly a fortnight after the metamorphosis, when the sexes are ready to pair.

Certain species of Neurothemis present, according to Brauer477 a curious case of dimorphism, some of the females having their wings netted in the usual manner; whilst other females have them "very richly netted as in the males of the same species." Brauer "explains the phenomenon on Darwinian principles by the supposition that the close netting of the veins is a secondary sexual character in the males." This latter character is generally developed in the males alone, but being, like every other masculine character, latent in the female, is occasionally developed in them. We have here an illustration of the manner in which the two sexes of many animals have probably come to resemble each other, namely by variations first appearing in the males, being preserved in them, and then transmitted to and developed in the females; but in this particular genus a complete transference is occasionally and abruptly effected. Mr. MacLachlan informs me of another case of dimorphism occurring in several species of Agrion in which a certain number of individuals are found of an orange colour, and these are 364invariably females. This is probably a case of reversion, for in the true Libellulæ, when the sexes differ in colour, the females are always orange or yellow, so that supposing Agrion to be descended from some primordial form having the characteristic sexual colours of the typical Libellulæ, it would not be surprising that a tendency to vary in this manner should occur in the females alone.

Although many dragon-flies are such large, powerful, and fierce insects, the males have not been observed by Mr. MacLachlan to fight together, except, as he believes, in the case of some of the smaller species of Agrion. In another very distinct group in this Order, namely in the Termites or white ants, both sexes at the time of swarming may be seen running about, "the male after the female, sometimes two chasing one female, and contending with great eagerness who shall win the prize."478

Order, Hymenoptera.—That inimitable observer, M. Fabre,479 in describing the habits of Cerceris, a wasp-like insect, remarks that "fights frequently ensue between the males for the possession of some particular female, who sits an apparently unconcerned beholder of the struggle for supremacy, and when the victory is decided, quietly flies away in company with the conqueror." Westwood480 says that the males of one of the saw-flies (Tenthredinæ) "have been found fighting together, with their mandibles locked." As M. Fabre speaks of the males of Cerceris striving to obtain a particular female, it may be well to bear in 365mind that insects belonging to this Order have the power of recognising each other after long intervals of time, and are deeply attached. For instance, Pierre Huber, whose accuracy no one doubts, separated some ants, and when after an interval of four months they met others which had formerly belonged to the same community, they mutually recognised and caressed each other with their antennæ. Had they been strangers they would have fought together. Again, when two communities engage in a battle, the ants on the same side in the general confusion sometimes attack each other, but they soon perceive their mistake, and the one ant soothes the other.481

In this Order slight differences in colour, according to sex, are common, but conspicuous differences are rare except in the family of Bees; yet both sexes of certain groups are so brilliantly coloured—for instance in Chrysis, in which vermilion and metallic greens prevail—that we are tempted to attribute the result to sexual selection. In the Ichneumonidæ, according to Mr. Walsh,482 the males are almost universally lighter coloured than the females. On the other hand, in the Tenthredinidæ the males are generally darker than the females. In the Siricidæ the sexes frequently differ; thus the male of Sirex juvencus is banded with orange, whilst the female is dark purple; but it is difficult to say which sex is the most ornamented. In Tremex columbæ the female is much brighter coloured than the male. With ants, as I am informed by Mr. F. Smith, the males of several species are black, the females being testaceous. In the family of Bees, especially in 366the solitary species, as I hear from the same distinguished entomologist, the sexes often differ in colour. The males are generally the brightest, and in Bombus as well as in Apathus, much more variable in colour than the females. In Anthophora retusa the male is of a rich fulvous-brown, whilst the female is quite black: so are the females of several species of Xylocopa, the males being bright yellow. In an Australian bee (Lestis bombylans), the female is of an extremely brilliant steel-blue, sometimes tinted with vivid green; the male being of a bright brassy colour clothed with rich fulvous pubescence. As in this group the females are provided with excellent defensive weapons in their stings, it is not probable that they have come to differ in colour from the males for the sake of protection.

Mutilla Europæa emits a stridulating noise; and according to Goureau483 both sexes have this power. He attributes the sound to the friction of the third and preceding abdominal segments; and I find that these surfaces are marked with very fine concentric ridges, but so is the projecting thoracic collar, on which the head articulates; and this collar, when scratched with the point of a needle, emits the proper sound. It is rather surprising that both sexes should have the power of stridulating, as the male is winged and the female wingless. It is notorious that Bees express certain emotions, as of anger, by the tone of their humming, as do some dipterous insects; but I have not referred to these sounds, as they are not known to be in any way connected with the act of courtship.

Order, Coleoptera (Beetles).—Many beetles are coloured so as to resemble the surfaces which they 367habitually frequent. Other species are ornamented with gorgeous metallic tints,—for instance, many Carabidæ, which live on the ground and have the power of defending themselves by an intensely acrid secretion,—the splendid diamond-beetles which are protected by an extremely hard covering,—many species of Chrysomela, such as C. cerealis, a large species beautifully striped with various colours, and in Britain confined to the bare summit of Snowdon,—and a host of other species. These splendid colours, which are often arranged in stripes, spots, crosses and other elegant patterns, can hardly be beneficial, as a protection, except in the case of some flower-feeding species; and we cannot believe that they are purposeless. Hence the suspicion arises, that they serve as a sexual attraction; but we have no evidence on this head, for the sexes rarely differ in colour. Blind beetles, which cannot of course behold each other's beauty, never exhibit, as I hear from Mr. Waterhouse, jun., bright colours, though they often have polished coats: but the explanation of their obscurity may be that blind insects inhabit caves and other obscure stations.

Some Longicorns, however, especially certain Prionidæ, offer an exception to the common rule that the sexes of beetles do not differ in colour. Most of these insects are large and splendidly coloured. The males in the genus Pyrodes,484 as I saw in Mr. Bates' collection, are 368generally redder but rather duller than the females, the latter being coloured of a more or less splendid golden green. On the other hand, in one species the male is golden-green, the female being richly tinted with red and purple. In the genus Esmeralda the sexes differ so greatly in colour that they have been ranked as distinct species: in one species both are of a beautiful shining green, but the male has a red thorax. On the whole, as far as I could judge, the females of those Prionidæ, in which the sexes differ, are coloured more richly than the males; and this does not accord with the common rule in regard to colour when acquired through sexual selection.

370

A most remarkable distinction between the sexes of many beetles is presented by the great horns which rise from the head, thorax, or clypeus of the males; and in some few cases from the under surface of the body. These horns, in the great family of the Lamellicorns, resemble those of various quadrupeds, such as stags, rhinoceroses, &c., and are wonderful both from their size and diversified shapes. Instead of describing them, I have given figures of the males and females of some of the more remarkable forms. (Figs. 15 to 19.) The females generally exhibit rudiments of the horns in the form of small knobs or ridges; but some are destitute of even a rudiment. On the other hand, the horns are nearly as well developed in the female as in the male of Phanæus lancifer; and only a little less well developed in the females of some other species of the same genus and of Copris. In the several subdivisions of the family, the differences in structure of the horns do not run parallel, as I am informed by Mr. Bates, with their more important and characteristic differences; thus within the same natural section of the genus Onthophagus, there are species which have either a single cephalic horn, or two distinct horns.

In almost all cases, the horns are remarkable from their excessive variability; so that a graduated series can be formed, from the most highly developed males to others so degenerate that they can barely be distinguished from the females. Mr. Walsh[485] found that in Phanæus carnifex the horns were thrice as long in some males as in others. Mr. Bates, after examining above a hundred males of Onthophagus rangifer (fig. 19), thought that he had at last discovered a species in 371which the horns did not vary; but further research proved the contrary.

The extraordinary size of the horns, and their widely different structure in closely-allied forms, indicate that they have been formed for some important purpose; but their excessive variability in the males of the same species leads to the inference that this purpose cannot be of a definite nature. The horns do not show marks of friction, as if used for any ordinary work. Some authors suppose[486] that as the males wander much more than the females, they require horns as a defence against their enemies; but in many cases the horns do not seem well adapted for defence, as they are not sharp. The most obvious conjecture is that they are used by the males for fighting together; but they have never been observed to fight; nor could Mr. Bates, after a careful examination of numerous species, find any sufficient evidence in their mutilated or broken condition of their having been thus used. If the males had been habitual fighters, their size would probably have been increased through sexual selection, so as to have exceeded that of the female; but Mr. Bates, after comparing the two sexes in above a hundred species of the Copridæ, does not find in well-developed individuals any marked difference in this respect. There is, moreover, one beetle, belonging to the same great division of the Lamellicorns, namely Lethrus, the males of which are known to fight, but they are not provided with horns, though their mandibles are much larger than those of the female.

The conclusion, which best agrees with the fact of the horns having been so immensely yet not fixedly developed,—as shewn by their extreme variability in 372the same species and by their extreme diversity in closely-allied species—is that they have been acquired as ornaments. This view will at first appear extremely improbable; but we shall hereafter find with many animals, standing much higher in the scale, namely fishes, amphibians, reptiles and birds, that various kinds of crests, knobs, horns and combs have been developed apparently for this sole purpose.

Onitis furcifer, male, viewed from beneath.with a great fork or pair of horns on the lower surface of the thorax. This situation seems extremely ill adapted for the display of these projections, and they may be of some real service; but no use can at present be assigned to them. It is a highly remarkable fact, that although the males do not exhibit even a trace of horns on the upper surface of the body, yet in the females a rudiment of a single horn on the head (fig. 21, a), and of a crest (b) on the thorax, are plainly visible. That the slight thoracic crest in the female is a rudiment of a projection proper to the male, though entirely absent in the male of this particular species, is clear: for the female of Bubas bison (a form373 which comes next to Onitis) has a similar slight crest on the thorax, and the male has in the same situation a great projection. So again there can be no doubt that the little point (a) on the head of the female Onitis furcifer, as well of the females of two or three allied species, is a rudimentary representative of the cephalic horn, which is common to the males of so many lamellicorn beetles, as in Phanæus, fig. 17. The males indeed of some unnamed beetles in the British Museum, which are believed actually to belong to the genus Onitis, are furnished with a similar horn. The remarkable nature of this case will be best perceived by an illustration: the Ruminant quadrupeds run parallel with the lamellicorn beetles, in some females possessing horns as large as those of the male, in others having them much smaller, or existing as mere rudiments (though this is as rare with ruminants as it is common with Lamellicorns), or in having none at all. Now if a new species of deer or sheep were discovered with the female bearing distinct rudiments of horns, whilst the head of the male was absolutely smooth, we should have a case like that of Onitis furcifer.

Fig. 21. Left-hand figure, male of Onitis furcifer, viewed laterally. Right-hand figure, female.   a. Rudiment of cephalic horn.   b. Trace of thoracic horn or crest.

In this case the old belief of rudiments having been created to complete the scheme of nature is so far from holding good, that all ordinary rules are completely broken through. The view which seems the most probable is that some early progenitor of Onitis acquired, like other Lamellicorns, horns on the head and thorax, and then transferred them, in a rudimentary condition, as with so many existing species, to the female, by whom they have ever since been retained. The subsequent loss of the horns by the male may have resulted through the principle of compensation from the development of the projections on the lower surface, whilst the female has not been thus affected, as she is not furnished with374 these projections, and consequently has retained the rudiments of the horns on the upper surface. Although this view is supported by the case of Bledius immediately to be given, yet the projections on the lower surface differ greatly in structure and development in the males of the several species of Onitis, and are even rudimentary in some; nevertheless the upper surface in all these species is quite destitute of horns. As secondary sexual characters are so eminently variable, it is possible that the projections on the lower surface may have been first acquired by some progenitor of Onitis and produced their effect through compensation, and then have been in certain cases almost completely lost.

All the cases hitherto given refer to the Lamellicorns, but the males of some few other beetles, belonging to two widely distinct groups, namely, the Curculionidæ and Staphylinidæ, are furnished with horns,—in the former on the lower surface of the body,487 in the latter on the upper surface of the head and thorax. In the Staphylinidæ the horns of the males in the same species are extraordinarily variable, just as we have seen with the Lamellicorns. In Siagonium we have a case of dimorphism, for the males can be divided into two sets, differing greatly in the size of their bodies, and in the development of their horns, without any intermediate gradations. In a species of Bledius (fig. 22), also belonging to the Staphylinidæ, male specimens can be found in the same locality, as 375Professor Westwood states, "in which the central horn of the thorax is very large, but the horns of the head quite rudimental; and others, in which the thoracic horn is much shorter, whilst the protuberances on the head are long."488 Here, then, we apparently have an instance of compensation of growth, which throws light on the curious case just given of the loss of the upper horns by the males of Onitis furcifer.

Law of Battle.—Some male beetles, which seem ill fitted for fighting, nevertheless engage in conflicts for the possession of the females. Mr. Wallace489 saw two males ofLeptorhynchus angustatus, a linear beetle with a much elongated rostrum, "fighting for a female, who stood close by busy at her boring. They pushed at each other with their rostra, and clawed and thumped, apparently in the greatest rage." The smaller male, however, "soon ran away, acknowledging himself vanquished." In some few cases the males are well adapted for fighting, by possessing great toothed mandibles, much larger than those of the females. This is the case with the common stag-beetle (Lucanus cervus), the males of which emerge from the pupal state about a week before the other sex, so that several may often be seen pursuing the same female. At this period they engage in fierce conflicts. When Mr. A. H. Davis[490] enclosed two males with one female in a box, the larger male severely pinched the smaller one, until he resigned his pretensions. A friend informs me 376that when a boy he often put the males together to see them fight, and he noticed that they were much bolder and fiercer than the females, as is well known to be the case with the higher animals. The males would seize hold of his finger, if held in front, but not so the females. With many of the Lucanidæ, as well as with the above-mentioned Leptorhynchus, the males are larger and more powerful insects than the females. The two sexes of Lethrus cephalotes (one of the Lamellicorns) inhabit the same burrow; and the male has larger mandibles than the female. If, during the breeding-season, a strange male attempts to enter the burrow, he is attacked; the female does not remain passive, but closes the mouth of the burrow, and encourages her mate by continually pushing him on from behind. The action does not cease until the aggressor is killed or runs away.[491] The two sexes of another lamellicorn beetle, the Ateuchus cicatricosus live in pairs, and seem much attached to each other; the male excites the female to roll the balls of dung in which the ova are deposited; and if she is removed, he becomes much agitated. If the male is removed, the female ceases all work, and as M. Brulerie[492] believes, would remain on the spot until she died.

The great mandibles of the male Lucanidæ are extremely variable both in size and structure, and in this respect resemble the horns on the head and thorax of many male Lamellicorns and Staphylinidæ. A perfect series can be formed from the best-provided to the worst-provided or degenerate males. Although the mandibles of the common stag-beetle, and probably of 377many other species, are used as efficient weapons for fighting, it is doubtful whether their great size can

Fig. 23. Chiasognathus grantii, reduced. Upper figure, male; lower figure, female.thus be accounted for. We have seen that with the Lucanus elaphus of N. America they are used for seizing the female. As they are so conspicuous and so elegantly branched, the suspicion has sometimes crossed my mind that they may be serviceable to the males as an ornament, in the same manner as the horns on the head and thorax of the various above described species. The male Chiasognathus grantii of S. Chile—a splendid beetle belonging to the same family—has enormously-developed mandibles (fig. 23); he is bold and pugnacious; when threatened on any side he faces round, opening his great jaws, and at the same time stridulating loudly; but the mandibles were not strong enough to pinch my finger so as to cause actual pain.

Sexual selection, which implies the possession of considerable perceptive powers and of strong passions, seems to have been more effective with the Lamellicorns than with any other family of the Coleoptera or beetles. With some species the males are provided with weapons for fighting; some live in pairs and show mutual affection;378 many have the power of stridulating when excited; many are furnished with the most extraordinary horns, apparently for the sake of ornament; some which are diurnal in their habits are gorgeously coloured; and, lastly, several of the largest beetles in the world belong to this family, which was placed by Linnæus and Fabricius at the head of the Order of the Coleoptera.493

Stridulating organs.—Beetles belonging to many and widely distinct families possess these organs. The sound can sometimes be heard at the distance of several feet or even yards,494 but is not comparable with that produced by the Orthoptera. The part which may be called the rasp generally consists of a narrow slightly-raised surface, crossed by very fine, parallel ribs, sometimes so fine as to cause iridescent colours, and having a very elegant appearance under the microscope. In some cases, for instance, with Typhæus, it could be plainly seen that extremely minute, bristly, scale-like prominences, which cover the whole surrounding surface in approximately parallel lines, give rise to the ribs of the rasp by becoming confluent and straight, and at the same time more prominent and smooth. A hard ridge on any adjoining part of the body, which in some cases is specially modified for the purpose, serves as the scraper for the rasp. The scraper is rapidly moved across the rasp, or conversely the rasp across the scraper.

These organs are situated in widely different positions. In the carrion-beetles (Necrophorus) two parallel rasps (r, fig. 24) stand on the dorsal surface of the fifth abdominal segment, each rasp being crossed, as described by Landois,495 by from 126 to 140 fine ribs. These 379ribs are scraped by the posterior margins of the elytra, a small portion of which projects beyond the general outline. In many Crioceridæ, and inClythra 4-punctata (one of the Chrysomelidæ), and in some Tenebrionidæ, &c.,496 the rasp is seated on the dorsal apex of the abdomen, on the pygidium or pro-pygidium, and is scraped as above by the elytra. In Heterocerus, which belongs to another family, the rasps are placed on the sides of the first abdominal segment, and are scraped by ridges on the femora.497 In certain Curculionidæ and Carabidæ,498 the parts are completely reversed in position, 380for the rasps are seated on the inferior surface of the elytra, near their apices, or along their outer margins, and the edges of the abdominal segments serve as the scrapers. In Pelobius hermanni (one of Dytiscidæ or water-beetles) a strong ridge runs parallel and near to the sutural margin of the elytra, and is crossed by ribs, coarse in the middle part, but becoming gradually finer at both ends, especially at the upper end; when this insect is held under water or in the air, a stridulating noise is produced by scraping the extreme horny margin of the abdomen against the rasp. In a great number of long-horned beetles (Longicornia) the organs are altogether differently situated, the rasp being on the meso-thorax, which is rubbed against the pro-thorax; Landois counted 238 very fine ribs on the rasp of Cerambyx heros.

Many Lamellicorns have the power of stridulating, and the organs differ greatly in position. Some species r. Rasp. c. Coxa. f. Femur. t. Tibia.tr. Tarsi.stridulate very loudly, so that when Mr. F. Smith caught a Trox sabulosus, a gamekeeper who stood by thought that he had caught a mouse; but I failed to discover the proper organs in this beetle. In Geotrupes and Typhæus a narrow ridge runs obliquely across (r, fig. 25) the coxa of each hind-leg, having in G. stercorarius 84 ribs, which are scraped by a specially-projecting part of one of the abdominal segments. In the nearly allied Copris lunaris, an excessively narrow fine rasp runs along the sutural margin of the elytra, with another short rasp near the basal outer margin; but in some other Coprini381 the rasp is seated, according to Leconte,499 on the dorsal surface of the abdomen. In Oryctes it is seated on the pro-pygidium, and in some other Dynastini, according to the same entomologist, on the under surface of the elytra. Lastly, Westring states that in Omaloplia brunnea the rasp is placed on the pro-sternum, and the scraper on the meta-sternum, the parts thus occupying the under surface of the body, instead of the upper surface as in the Longicorns.

We thus see that the stridulating organs in the different coleopterous families are wonderfully diversified in position, but not much in structure. Within the same family some species are provided with these organs, and some are quite destitute of them. This diversity is intelligible, if we suppose that originally various species made a shuffling or hissing noise by the rubbing together of the hard and rough parts of their bodies which were in contact; and that from the noise thus produced being in some way useful, the rough surfaces were gradually developed into regular stridulating organs. Some beetles as they move, now produce, either intentionally or unintentionally, a shuffling noise, without possessing any proper organs for the purpose. Mr. Wallace informs me that the Euchirus longimanus (a Lamellicorn, with the anterior legs wonderfully elongated in the male) "makes, whilst moving, a low hissing sound by the protrusion and contraction of the abdomen; and when seized it produces a grating sound by rubbing its hind-legs against the edges of the elytra." The hissing sound is clearly due to a narrow rasp running along the sutural margin of each elytron; and I could likewise make the grating 382sound by rubbing the shagreened surface of the femur against the granulated margin of the corresponding elytron; but I could not here detect any proper rasp; nor is it likely that I could have overlooked it in so large an insect. After examining Cychrus and reading what Westring has written in his two papers about this beetle, it seems very doubtful whether it possesses any true rasp, though it has the power of emitting a sound.

From the analogy of the Orthoptera and Homoptera, I expected to find that the stridulating organs in the Coleoptera differed according to sex; but Landois, who has carefully examined several species, observed no such difference; nor did Westring; nor did Mr. G. R. Crotch in preparing the numerous specimens which he had the kindness to send me for examination. Any slight sexual difference, however, would be difficult to detect, on account of the great variability of these organs. Thus in the first pair of the Necrophorus humator and of the Pelobius which I examined, the rasp was considerably larger in the male than in the female; but not so with succeeding specimens. In Geotrupes stercorariusthe rasp appeared to me thicker, opaquer, and more prominent in three males than in the same number of females; consequently my son, Mr. F. Darwin, in order to discover whether the sexes differed in their power of stridulating, collected 57 living specimens, which he separated into two lots, according as they made, when held in the same manner, a greater or lesser noise. He then examined their sexes, but found that the males were very nearly in the same proportion to the females in both lots. Mr. F. Smith has kept alive numerous specimens of Mononychus pseudacori (Curculionidæ), and is satisfied that both sexes stridulate, and apparently in an equal degree.

Nevertheless the power of stridulating is certainly a383 sexual character in some few Coleoptera. Mr. Crotch has discovered that the males alone of two species of Heliopathes (Tenebrionidæ) possess stridulating organs. I examined five males of H. gibbus, and in all these there was a well-developed rasp, partially divided into two, on the dorsal surface of the terminal abdominal segment; whilst in the same number of females there was not even a rudiment of the rasp, the membrane of this segment being transparent and much thinner than in the male. In H. cribratostriatus the male has a similar rasp, excepting that it is not partially divided into two portions, and the female is completely destitute of this organ; but in addition the male has on the apical margins of the elytra, on each side of the suture, three or four short longitudinal ridges, which are crossed by extremely fine ribs, parallel to and resembling those on the abdominal rasp; whether these ridges serve as an independent rasp, or as a scraper for the abdominal rasp, I could not decide: the female exhibits no trace of this latter structure.

Again, in three species of the Lamellicorn genus Oryctes, we have a nearly parallel case. In the females of O. gryphus and nasicornis the ribs on the rasp of the pro-pygidium are less continuous and less distinct than in the males; but the chief difference is that the whole upper surface of this segment, when held in the proper light, is seen to be clothed with hairs, which are absent or are represented by excessively fine down in the males. It should be noticed that in all Coleoptera the effective part of the rasp is destitute of hairs. In O. senegalensis the difference between the sexes is more strongly marked, and this is best seen when the proper segment is cleaned and viewed as a transparent object. In the female the whole surface is covered with little separate crests, bearing spines; whilst in the male these crests384 become, in proceeding towards the apex, more and more confluent, regular, and naked; so that three-fourths of the segment is covered with extremely fine parallel ribs, which are quite absent in the female. In the females, however, of all three species of Oryctes, when the abdomen of a softened specimen is pushed backwards and forwards, a slight grating or stridulating sound can be produced.

In the case of the Heliopathes and Oryctes there can hardly be a doubt that the males stridulate in order to call or to excite the females; but with most beetles the stridulation apparently serves both sexes as a mutual call. This view is not rendered improbable from beetles stridulating under various emotions; we know that birds use their voices for many purposes besides singing to their mates. The great Chiasognathus stridulates in anger or defiance; many species do the same from distress or fear, when held so that they cannot escape; Messrs. Wollaston and Crotch were able, by striking the hollow stems of trees in the Canary Islands, to discover the presence of beetles belonging to the genus Acalles by their stridulation. Lastly the male Ateuchus stridulates to encourage the female in her work, and from distress when she is removed.500 Some naturalists believe that beetles make this noise to frighten away their enemies; but I cannot think that the quadrupeds and birds which are able to devour the larger beetles with their extremely hard coats, would be frightened by so slight a grating sound. The belief that the stridulation serves as a sexual call is supported by the fact that death-ticks (Anobium tesselatum) are well known to answer each other's ticking, or, as I have 385myself observed, a tapping noise artificially made; and Mr. Doubleday informs me that he has twice or thrice observed a female ticking,501and in the course of an hour or two has found her united with a male, and on one occasion surrounded by several males. Finally, it seems probable that the two sexes of many kinds of beetles were at first enabled to find each other by the slight shuffling noise produced by the rubbing together of the adjoining parts of their hard bodies; and that as the males or females which made the greatest noise succeeded best in finding partners, the rugosities on various parts of their bodies were gradually developed by means of sexual selection into true stridulating organs.

CHAPTER XI.

Insects, continued.—Order Lepidoptera.

Courtship of butterflies—Battles—Ticking noise—Colours common to both sexes, or more brilliant in the males—Examples—Not due to the direct action of the conditions of life—Colours adapted for protection—Colours of moths—Display—Perceptive powers of the Lepidoptera—Variability—Causes of the difference in colour between the males and females—Mimickry, female butterflies more brilliantly coloured than the males—Bright colours of caterpillars—Summary and concluding remarks on the secondary sexual characters of insects—Birds and insects compared.

In this great Order the most interesting point for us is the difference in colour between the sexes of the same species, and between the distinct species of the same genus. Nearly the whole of the following chapter will be devoted to this subject; but I will first make a few remarks on one or two other points. Several males may often be seen pursuing and crowding round the same female. Their courtship appears to be a prolonged affair, for I have frequently watched one or more males pirouetting round a female until I became tired, without seeing the end of the courtship. Although butterflies are such weak and fragile creatures, they are pugnacious, and an Emperor butterfly502 has been captured with the tips of its wings broken from a conflict with another male. Mr. Collingwood in speaking of the frequent battles 387between the butterflies of Borneo says, "They whirl round each other with the greatest rapidity, and appear to be incited by the greatest ferocity." One case is known of a butterfly, namely the Ageronia feronia, which makes a noise like that produced by a toothed wheel passing under a spring catch, and which could be heard at the distance of several yards. At Rio de Janeiro this sound was noticed by me, only when two were chasing each other in an irregular course, so that it is probably made during the courtship of the sexes; but I neglected to attend to this point.503

Every one has admired the extreme beauty of many butterflies and of some moths; and we are led to ask, how has this beauty been acquired? Have their colours and diversified patterns simply resulted from the direct action of the physical conditions to which these insects have been exposed, without any benefit being thus derived? Or have successive variations been accumulated and determined either as a protection or for some unknown purpose, or that one sex might be rendered attractive to the other? And, again, what is the meaning of the colours being widely different in the males and females of certain species, and alike in the two sexes of other species? Before attempting to answer these questions a body of facts must be given.

With most of our English butterflies, both those which are beautiful, such as the admiral, peacock, and painted lady (Vanessæ), and those which are plain-coloured, such as the meadow-browns (Hipparchiæ), the sexes are alike. This is also the case with the magnificent Heliconidæ and Danaidæ of the tropics. But in certain 388other tropical groups, and with some of our English butterflies, as the purple emperor, orange-tip, &c. (Apatura Iris and Anthocharis cardamines), the sexes differ either greatly or slightly in colour. No language suffices to describe the splendour of the males of some tropical species. Even within the same genus we often find species presenting an extraordinary difference between the sexes, whilst others have their sexes closely alike. Thus in the South American genus Epicalia, Mr. Bates, to whom I am much indebted for most of the following facts and for looking over this whole discussion, informs me that he knows twelve species, the two sexes of which haunt the same stations (and this is not always the case with butterflies), and therefore cannot have been differently affected by external conditions504. In nine of these species the males rank amongst the most brilliant of all butterflies, and differ so greatly from the comparatively plain females that they were formerly placed in distinct genera. The females of these nine species resemble each other in their general type of coloration, and likewise resemble both sexes in several allied genera, found in various parts of the world. Hence in accordance with the descent-theory we may infer that these nine species, and probably all the others of the genus, are descended from an ancestral form which was coloured in nearly the same manner. In the tenth species the female still retains the same general colouring, but the male resembles her, so that he is coloured in a much less gaudy and contrasted manner than the males of the previous species. In the eleventh and twelfth species, the females depart from the type of colouring which 389is usual with their sex in this genus, for they are gaily decorated in nearly the same manner as the males, but in a somewhat less degree. Hence in these two species the bright colours of the males seem to have been transferred to the females; whilst the male of the tenth species has either retained or recovered the plain colours of the female as well as of the parent-form of the genus; the two sexes being thus rendered in both cases, though in an opposite manner, nearly alike. In the allied genus Eubagis, both sexes of some of the species are plain-coloured and nearly alike; whilst with the greater number the males are decorated with beautiful metallic tints, in a diversified manner, and differ much from their females. The females throughout the genus retain the same general style of colouring, so that they commonly resemble each other much more closely than they resemble their own proper males.

In the genus Papilio, all the species of the Æneas group are remarkable for their conspicuous and strongly contrasted colours, and they illustrate the frequent tendency to gradation in the amount of difference between the sexes. In a few species, for instance in P. ascanius, the males and females are alike; in others the males are a little or very much more superbly coloured than the females. The genus Junonia allied to our Vanessæ offers a nearly parallel case, for although the sexes of most of the species resemble each other and are destitute of rich colours, yet in certain species, as in J. œnone, the male is rather more brightly coloured than the female, and in a few (for instance J. andremiaja) the male is so different from the female that he might be mistaken for an entirely distinct species.

Another striking case was pointed out to me in the British museum by Mr. A. Butler, namely one of the Tropical American Theclæ, in which both sexes390 are nearly alike and wonderfully splendid; in another, the male is coloured in a similarly gorgeous manner, whilst the whole upper surface of the female is of a dull uniform brown. Our common little English blue butterflies of the genus Lycæna, illustrate the various differences in colour between the sexes, almost as well, though not in so striking a manner, as the above exotic genera. In Lycæna agestis both sexes have wings of a brown colour, bordered with small ocellated orange spots, and are consequently alike. In L. œgon the wings of the male are of a fine blue, bordered with black; whilst the wings of the female are brown, with a similar border, and closely resemble those of L. agestis. Lastly, in L. arion both sexes are of a blue colour and nearly alike, though in the female the edges of the wings are rather duskier, with the black spots plainer; and in a bright blue Indian species both sexes are still more closely alike.

I have given the foregoing cases in some detail in order to shew, in the first place, that when the sexes of butterflies differ, the male as a general rule is the most beautiful, and departs most from the usual type of colouring of the group to which the species belongs. Hence in most groups the females of the several species resemble each other much more closely than do the males. In some exceptional cases, however, to which I shall hereafter allude, the females are coloured more splendidly than the males. In the second place these cases have been given to bring clearly before the mind that within the same genus, the two sexes frequently present every gradation from no difference in colour to so great a difference that it was long before the two were placed by entomologists in the same genus. In the third place, we have seen that when the sexes nearly resemble each other, this apparently may be due either to the391 male having transferred his colours to the female, or to the male having retained, or perhaps recovered, the primordial colours of the genus to which the species belongs. It also deserves notice that in those groups in which the sexes present any difference of colour, the females usually resemble the males to a certain extent, so that when the males are beautiful to an extraordinary degree, the females almost invariably exhibit some degree of beauty. From the numerous cases of gradation in the amount of difference between the sexes, and from the prevalence of the same general type of coloration throughout the whole of the same group, we may conclude that the causes, whatever they may be, which have determined the brilliant colouring of the males alone of some species, and of both sexes in a more or less equal degree of other species, have generally been the same.

As so many gorgeous butterflies inhabit the tropics, it has often been supposed that they owe their colours to the great heat and moisture of these zones; but Mr. Bates505 has shewn by the comparison of various closely-allied groups of insects from the temperate and tropical regions, that this view cannot be maintained; and the evidence becomes conclusive when brilliantly-coloured males and plain-coloured females of the same species inhabit the same district, feed on the same food, and follow exactly the same habits of life. Even when the sexes resemble each other, we can hardly believe that their brilliant and beautifully-arranged colours are the purposeless result of the nature of the tissues, and the action of the surrounding conditions.

With animals of all kinds, whenever colour has been modified for some special purpose, this has been, as far 392as we can judge, either for protection or as an attraction between the sexes. With many species of butterflies the upper surfaces of the wings are obscurely coloured, and this in all probability leads to their escaping observation and danger. But butterflies when at rest would be particularly liable to be attacked by their enemies; and almost all the kinds when resting raise their wings vertically over their backs, so that the lower sides alone are exposed to view. Hence it is this side which in many cases is obviously coloured so as to imitate the surfaces on which these insects commonly rest. Dr. Rössler, I believe, first noticed the similarity of the closed wings of certain Vanessæ and other butterflies to the bark of trees. Many analogous and striking facts could be given. The most interesting one is that recorded by Mr. Wallace506 of a common Indian and Sumatran butterfly (Kallima), which disappears like magic when it settles in a bush; for it hides its head and antennæ between its closed wings, and these in form, colour, and veining cannot be distinguished from a withered leaf together with the footstalk. In some other cases the lower surfaces of the wings are brilliantly coloured, and yet are protective; thus in Thecla rubi the wings when closed are of an emerald green and resemble the young leaves of the bramble, on which this butterfly in the spring may often be seen seated.

Although the obscure tints of the upper or under surface of many butterflies no doubt serve to conceal them, yet we cannot possibly extend this view to the brilliant and conspicuous colours of many kinds, such as our admiral and peacock Vanessæ, our white 393cabbage-butterflies (Pieris), or the great swallow-tail Papilio which haunts the open fens—for these butterflies are thus rendered visible to every living creature. With these species both sexes are alike; but in the common brimstone butterfly (Gonepteryx rhamni), the male is of an intense yellow, whilst the female is much paler; and in the orange-tip (Anthocharis cardamines) the males alone have the bright orange tips to their wings. In these cases the males and females are equally conspicuous, and it is not credible that their difference in colour stands in any relation to ordinary protection. Nevertheless it is possible that the conspicuous colours of many species may be in an indirect manner beneficial, as will hereafter be explained, by leading their enemies at once to recognise them as unpalatable. Even in this case it does not certainly follow that their bright colours and beautiful patterns were acquired for this special purpose. In some other remarkable cases, beauty has been gained for the sake of protection, through the imitation of other beautiful species, which inhabit the same district and enjoy an immunity from attack by being in some way offensive to their enemies.

The female of our orange-tip butterfly, above referred to, and of an American species (Anth. genutia) probably shew us, as Mr. Walsh has remarked to me, the primordial colours of the parent-species of the genus; for both sexes of four or five widely-distributed species are coloured in nearly the same manner. We may infer here, as in several previous cases, that it is the males of Anth. cardamines and genutia which have departed from the usual type of colouring of their genus. In the Anth. sara from California, the orange-tips have become partially developed in the female; for her wings are tipped with reddish-orange, but paler than in the394 male, and slightly different in some other respects. In an allied Indian form, the Iphias glaucippe, the orange-tips are fully developed in both sexes. In this Iphias the under surface of the wings marvellously resembles, as pointed out to me by Mr. A. Butler, a pale-coloured leaf; and in our English orange-tip, the under surface resembles the flower-head of the wild parsley, on which it may be seen going to rest at night.507 The same reasoning power which compels us to believe that the lower surfaces have here been coloured for the sake of protection, leads us to deny that the wings have been tipped, especially when this character is confined to the males, with bright orange for the same purpose.

Turning now to Moths: most of these rest motionless with their wings depressed during the whole or greater part of the day; and the upper surfaces of their wings are often shaded and coloured in an admirable manner, as Mr. Wallace has remarked, for escaping detection. With most of the Bombycidæ and Noctuidæ,508 when at rest, the front-wings overlap and conceal the hind-wings; so that the latter might be brightly coloured without much risk; and they are thus coloured in many species of both families. During the act of flight, moths would often be able to escape from their enemies; nevertheless, as the hind-wings are then fully exposed to view, their bright colours must generally have been acquired at the cost of some little risk. But the following fact shews us how cautious we ought to be in drawing conclusions on this head. The common yellow under-wings 395(Triphaena) often fly about during the day or early evening, and are then conspicuous from the colour of their hind-wings. It would naturally be thought that this would be a source of danger; but Mr. J. Jenner Weir believes that it actually serves them as a means of escape, for birds strike at these brightly coloured and fragile surfaces, instead of at the body. For instance, Mr. Weir turned into his aviary a vigorous specimen of Triphaena pronuba, which was instantly pursued by a robin; but the bird's attention being caught by the coloured wings, the moth was not captured until after about fifty attempts, and small portions of the wings were repeatedly broken off. He tried the same experiment, in the open air, with a T. fimbria and swallow; but the large size of this moth probably interfered with its capture.509 We are thus reminded of a statement made by Mr. Wallace,510 namely, that in the Brazilian forests and Malayan islands, many common and highly-decorated butterflies are weak flyers, though furnished with a broad expanse of wings; and they "are often captured with pierced and broken wings, as if they had been seized by birds, from which they had escaped: if the wings had been much smaller in proportion to the body, it seems probable that the insect would more frequently have been struck or pierced in a vital part, and thus the increased expanse of the wings may have been indirectly beneficial."

Display.—The bright colours of butterflies and of some moths are specially arranged for display, whether or not they serve in addition as a protection. Bright 396colours would not be visible during the night; and there can be no doubt that moths, taken as a body, are much less gaily decorated than butterflies, all of which are diurnal in their habits. But the moths in certain families, such as the Zygænidæ, various Sphingidæ, Uraniidæ, some Arctiidæ and Saturniidæ, fly about during the day or early evening, and many of these are extremely beautiful, being far more brightly coloured than the strictly nocturnal kinds. A few exceptional cases, however, of brightly-coloured nocturnal species have been recorded.511

There is evidence of another kind in regard to display. Butterflies, as before remarked, elevate their wings when at rest, and whilst basking in the sunshine often alternately raise and depress them, thus exposing to full view both surfaces; and although the lower surface is often coloured in an obscure manner as a protection, yet in many species it is as highly coloured as the upper surface, and sometimes in a very different manner. In some tropical species the lower surface is even more brilliantly coloured than the upper.512 In one English fritillary, the Argynnis aglaia, the lower surface alone is ornamented with shining silver discs. Nevertheless, as a general rule, the upper surface, which is probably the most fully exposed, is coloured more brightly and in a more diversified manner than the lower. Hence the lower surface generally affords 397to entomologists the most useful character for detecting the affinities of the various species.

Now if we turn to the enormous group of moths, which do not habitually expose to full view the under surface of their wings, this side is very rarely, as I hear from Mr. Stainton, coloured more brightly than the upper side, or even with equal brightness. Some exceptions to the rule, either real or apparent, must be noticed, as that of Hypopira, specified by Mr. Wormald.513 Mr. R. Trimen informs me that in Guenée's great work, three moths are figured, in which the under surface is much the most brilliant. For instance, in the Australian Gastrophora the upper surface of the fore-wing is pale greyish-ochreous, while the lower surface is magnificently ornamented by an ocellus of cobalt-blue, placed in the midst of a black mark, surrounded by orange-yellow, and this by bluish-white. But the habits of these three moths are unknown; so that no explanation can be given of their unusual style of colouring. Mr. Trimen also informs me that the lower surface of the wings in certain other Geometræ514 and quadrifid Noctuæ are either more variegated or more brightly-coloured than the upper surface; but some of these species have the habit of "holding their wings quite erect over their backs, retaining them in this position for a considerable time," and thus exposing to view the under surface. Other species when settled on the ground or herbage have the habit of now and then suddenly and slightly lifting up their wings. Hence the lower surface of the wings being more brightly-coloured than the upper sur398face in certain moths is not so anomalous a circumstance as it at first appears. The Saturniidæ include some of the most beautiful of all moths, their wings being decorated, as in our British Emperor moth, with fine ocelli; and Mr. T. W. Wood515 observes that they resemble butterflies in some of their movements; "for instance, in the gentle waving up and down of the wings, as if for display, which is more characteristic of diurnal than of nocturnal Lepidoptera."

It is a singular fact that no British moths, nor as far as I can discover hardly any foreign species, which are brilliantly coloured, differ much in colour according to sex; though this is the case with many brilliant butterflies. The male, however, of one American moth, the Saturnia Io, is described as having its fore-wings deep yellow, curiously marked with purplish-red spots; whilst the wings of the female are purple-brown, marked with grey lines.516 The British moths which differ sexually in colour are all brown, or various tints of dull yellow, or nearly white. In several species the males are much darker than the females,517 and these belong to groups which generally fly about during the afternoon. On the other hand, in many genera, as Mr. Stainton informs me, 399the males have the hind-wings whiter than those of the female—of which, fact Agrotis exclamationis offers a good instance. The males are thus rendered more conspicuous than the females, whilst flying about in the dusk. In the Ghost Moth (Hepialus humuli) the difference is more strongly marked; the males being white and the females yellow with darker markings. It is difficult to conjecture what the meaning can be of these differences between the sexes in the shades of darkness or lightness; but we can hardly suppose that they are the result of mere variability with sexually-limited inheritance, independently of any benefit thus derived.

From the foregoing statements it is impossible to admit that the brilliant colours of butterflies and of some few moths, have commonly been acquired for the sake of protection. We have seen that their colours and elegant patterns are arranged and exhibited as if for display. Hence I am led to suppose that the females generally prefer, or are most excited by the more brilliant males; for on any other supposition the males would be ornamented, as far as we can see, for no purpose. We know that ants and certain lamellicorn beetles are capable of feeling an attachment for each other, and that ants recognise their fellows after an interval of several months. Hence there is no abstract improbability in the Lepidoptera, which probably stand nearly or quite as high in the scale as these insects, having sufficient mental capacity to admire bright colours. They certainly discover flowers by colour, and, as I have elsewhere shewn, the plants which are fertilised exclusively by the wind never have a conspicuously-coloured corolla. The Humming-bird Sphinx may often be seen to swoop down from a distance on a bunch of flowers in the midst of green foliage;400 and I have been assured by a friend, that these moths repeatedly visited flowers painted on the walls of a room in the South of France. The common white butterfly, as I hear from Mr. Doubleday, often flies down to a bit of paper on the ground, no doubt mistaking it for one of its own species. Mr. Collingwood518 in speaking of the difficulty of collecting certain butterflies in the Malay Archipelago, states that "a dead specimen pinned upon a conspicuous twig will often arrest an insect of the same species in its headlong flight, and bring it down within easy reach of the net, especially if it be of the opposite sex."

The courtship of butterflies is a prolonged affair. The males sometimes fight together in rivalry; and many may be seen pursuing or crowding round the same female. If, then, the females do not prefer one male to another, the pairing must be left to mere chance, and this does not appear to me a probable event. If, on the other hand, the females habitually, or even occasionally, prefer the more beautiful males, the colours of the latter will have been rendered brighter by degrees, and will have been transmitted to both sexes or to one sex, according to which law of inheritance prevailed. The process of sexual selection will have been much facilitated, if the conclusions arrived at from various kinds of evidence in the supplement to the ninth chapter can be trusted; namely that the males of many Lepidoptera, at least in the imago state, greatly exceed in number the females.

Some facts, however, are opposed to the belief that female butterflies prefer the more beautiful males; thus, as I have been assured by several observers, fresh females may frequently be seen paired with battered, faded or 401dingy males; but this is a circumstance which could hardly fail often to follow from the males emerging from their cocoons earlier than the females. With moths of the family of the Bombycidæ, the sexes pair immediately after assuming the imago state; for they cannot feed, owing to the rudimentary condition of their mouths. The females, as several entomologists have remarked to me, lie in an almost torpid state, and appear not to evince the least choice in regard to their partners, This is the case with the common silk-moth (B. mori), as I have been told by some continental and English breeders. Dr. Wallace, who has had such immense experience in breedingBombyx cynthia, is convinced that the females evince no choice or preference. He has kept above 300 of these moths living together, and has often found the most vigorous females mated with stunted males. The reverse apparently seldom occurs; for, as he believes, the more vigorous males pass over the weakly females, being attracted by those endowed with most vitality. Although we have been indirectly induced to believe that the females of many species prefer the more beautiful males, I have no reason to suspect, either with moths or butterflies, that the males are attracted by the beauty of the females. If the more beautiful females had been continually preferred, it is almost certain, from the colours of butterflies being so frequently transmitted to one sex alone, that the females would often have been rendered more beautiful than their male partners. But this does not occur except in a few instances; and these can be explained, as we shall presently see, on the principle of mimickry and protection.

As sexual selection primarily depends on variability, a few words must be added on this subject. In respect402 to colour there is no difficulty, as any number of highly variable Lepidoptera could be named. One good instance will suffice. Mr. Bates shewed me a whole series of specimens of Papilio sesostris and childrenæ; in the latter the males varied much in the extent of the beautifully enamelled green patch on the fore-wings, and in the size of the white mark, as well as of the splendid crimson stripe on the hind-wings; so that there was a great contrast between the most and least gaudy males. The male of Papilio sesostris, though a beautiful insect, is much less so than P. childrenæ. It likewise varies a little in the size of the green patch on the fore-wings, and in the occasional appearance of a small crimson stripe on the hind-wings, borrowed, as it would seem, from its own female; for the females of this and of many other species in the Æneas group possess this crimson stripe. Hence between the brightest specimens of P. sesostris and the least bright of P. childrenæ, there was but a small interval; and it was evident that as far as mere variability is concerned, there would be no difficulty in permanently increasing by means of selection the beauty of either species. The variability is here almost confined to the male sex; but Mr. Wallace and Mr. Bates have shewn519 that the females of some other species are extremely variable, the males being nearly constant. As I have before mentioned the Ghost Moth (Hepialus humuli) as one of the best instances in Britain of a difference in colour between the sexes of moths, it may be worth adding520 that in the Shetland 403Islands, males are frequently found which closely resemble the females. In a future chapter I shall have occasion to shew that the beautiful eye-like spots or ocelli, so common on the wings of many Lepidoptera, are eminently variable.

On the whole, although many serious objections may be urged, it seems probable that most of the species of Lepidoptera which are brilliantly coloured, owe their colours to sexual selection, excepting in certain cases, presently to be mentioned, in which conspicuous colours are beneficial as a protection. From the ardour of the male throughout the animal kingdom, he is generally willing to accept any female; and it is the female which usually exerts a choice. Hence if sexual selection has here acted, the male, when the sexes differ, ought to be the most brilliantly coloured; and this undoubtedly is the ordinary rule. When the sexes are brilliantly coloured and resemble each other, the characters acquired by the males appear to have been transmitted to both sexes. But will this explanation of the similarity and dissimilarity in colour between the sexes suffice?

The males and females of the same species of butterfly are known521 in several cases to inhabit different stations, the former commonly basking in the sunshine, the latter haunting gloomy forests. It is therefore possible that different conditions of life may have acted directly on the two sexes; but this is not probable,522 as in the adult state they are exposed during a very short period to different conditions; and the larvæ of both are exposed to the same conditions. Mr. Wallace believes 404that the less brilliant colours of the female have been specially gained in all or almost all cases for the sake of protection. On the contrary it seems to me more probable that the males alone, in the large majority of cases, have acquired their bright colours through sexual selection, the females having been but little modified. Consequently the females of distinct but allied species ought to resemble each other much more closely than do the males of the same species; and this is the general rule. The females thus approximately show us the primordial colouring of the parent-species of the group to which they belong. They have, however, almost always been modified to a certain extent by some of the successive steps of variation, through the accumulation of which the males were rendered beautiful, having been transferred to them. The males and females of allied though distinct species will also generally have been exposed during their prolonged larval state to different conditions, and may have been thus indirectly affected; though with the males any slight change of colour thus caused will often have been completely masked by the brilliant tints gained through sexual selection. When we treat of Birds, I shall have to discuss the whole question whether the differences in colour between the males and females have been in part specially gained by the latter as a protection; so that I will here only give unavoidable details.

In all cases when the more common form of equal inheritance by both sexes has prevailed, the selection of bright-coloured males would tend to make the females bright-coloured; and the selection of dull-coloured females would tend to make the males dull. If both processes were carried on simultaneously, they would tend to neutralise each other. As far as I can see, it would be extremely difficult to change through selection the405 one form of inheritance into the other. But by the selection of successive variations, which were from the first sexually limited in their transmission, there would not be the slightest difficulty in giving bright colours to the males alone, and at the same time or subsequently, dull colours to the females alone. In this latter manner female butterflies and moths may, as I fully admit, have been rendered inconspicuous for the sake of protection, and widely different from their males.

Mr. Wallace523 has argued with much force in favour of his view that when the sexes differ, the female has been specially modified for the sake of protection; and that this has been effected by one form of inheritance, namely, the transmission of characters to both sexes, having been changed through the agency of natural selection into the other form, namely, transmission to one sex. I was at first strongly inclined to accept this view; but the more I have studied the various classes throughout the animal kingdom, the less probable it has appeared. Mr. Wallace urges that both sexes of the Heliconidæ, Danaidæ, Acroeidæ are equally brilliant because both are protected from the attacks of birds and other enemies, by their offensive odour; but that in other groups, which do not possess this immunity, the females have been rendered inconspicuous, from having more need of protection than the males. This supposed difference in the "need of protection by the two sexes" is rather deceptive, and requires some discussion. It is obvious that brightly-coloured individuals, whether males or females, would equally attract, and obscurely-coloured individuals equally escape, the 406attention of their enemies. But we are concerned with the effects of the destruction or preservation of certain individuals of either sex, on the character of the race. With insects, after the male has fertilised the female, and after the latter has laid her eggs, the greater or less immunity from danger of either sex could not possibly have any effect on the offspring. Before the sexes have performed their proper functions, if they existed in equal numbers and if they strictly paired (all other circumstances being the same), the preservation of the males and females would be equally important for the existence of the species and for the character of the offspring. But with most animals, as is known to be the case with the domestic silk-moth, the male can fertilise two or three females; so that the destruction of the males would not be so injurious to the species as that of the females. On the other hand, Dr. Wallace believes that with moths the progeny from a second or third fertilisation is apt to be weakly, and therefore would not have so good chance of surviving. When the males exist in much greater numbers than the females, no doubt many males might be destroyed with impunity to the species; but I cannot see that the results of ordinary selection for the sake of protection would be influenced by the sexes existing in unequal numbers; for the same proportion of the more conspicuous individuals, whether males or females, would probably be destroyed. If indeed the males presented a greater range of variation in colour, the result would be different; but we need not here follow out such complex details. On the whole I cannot perceive that an inequality in the numbers of the two sexes would influence in any marked manner the effects of ordinary selection on the character of the offspring.

Female Lepidoptera require, as Mr. Wallace insists,407 some days to deposit their fertilised ova and to search for a proper place; during this period (whilst the life of the male was of no importance) the brighter-coloured females would be exposed to danger and would be liable to be destroyed. The duller-coloured females on the other hand would survive, and thus would influence, it might be thought, in a marked manner the character of the species,— either of both sexes or of one sex, according to which form of inheritance prevailed. But it must not be forgotten that the males emerge from the cocoon-state some days before the females, and during this period, whilst the unborn females were safe, the brighter-coloured males would be exposed to danger; so that ultimately both sexes would probably be exposed during a nearly equal length of time to danger, and the elimination of conspicuous colours would not be much more effective in the one than the other sex.

It is a more important consideration that female Lepidoptera, as Mr. Wallace remarks, and as is known to every collector, are generally slower flyers than the males. Consequently the latter, if exposed to greater danger from being conspicuously coloured, might be able to escape from their enemies, whilst the similarly-coloured females would be destroyed; and thus the females would have the most influence in modifying the colour of their progeny.

There is one other consideration: bright colours, as far as sexual selection is concerned, are commonly of no service to the females; so that if the latter varied in brightness, and the variations were sexually limited in their transmission, it would depend on mere chance whether the females had their bright colours increased; and this would tend throughout the Order to diminish the number of species with brightly-coloured females408 in comparison with the species having brightly-coloured males. On the other hand, as bright colours are supposed to be highly serviceable to the males in their love-struggles, the brighter males (as we shall see in the chapter on Birds) although exposed to rather greater danger, would on an average procreate a greater number of offspring than the duller males. In this case, if the variations were limited in their transmission to the male sex, the males alone would be rendered more brilliantly coloured; but if the variations were not thus limited, the preservation and augmentation of such variations would depend on whether more evil was caused to the species by the females being rendered conspicuous, than good to the males by certain individuals being successful over their rivals.

As there can hardly be a doubt that both sexes of many butterflies and moths have been rendered dull-coloured for the sake of protection, so it may have been with the females alone of some species in which successive variations towards dullness first appeared in the female sex and were from the first limited in their transmission to the same sex. If not thus limited, both sexes would become dull-coloured. We shall immediately see, when we treat of mimicry, that the females alone of certain butterflies have been rendered extremely beautiful for the sake of protection, without any of the successive protective variations having been transferred to the male, to whom they could not possibly have been in the least degree injurious, and therefore could not have been eliminated through natural selection. Whether in each particular species, in which the sexes differ in colour, it is the female which has been specially modified for the sake of protection; or whether it is the male which has been specially modified for the sake of sexual attraction, the 409 female having retained her primordial colouring only slightly changed through the agencies before alluded to; or whether again both sexes have been modified, the female for protection and the male for sexual attraction, can only be definitely decided when we know the life-history of each species.

Without distinct evidence, I am unwilling to admit that a double process of selection has long been going on with a multitude of species,—the males having been rendered more brilliant by beating their rivals; and the females more dull-coloured by having escaped from their enemies. We may take as an instance the common brimstone butterfly (Gonepteryx), which appears early in the spring before any other kind. The male of this species is of a far more intense yellow than the female, though she is almost equally conspicuous; and in this case it does not seem probable that she specially acquired her pale tints as a protection, though it is probable that the male acquired his bright colours as a sexual attraction. The female of Anthocharis cardamines does not possess the beautiful orange tips to her wings with which the male is ornamented; consequently she closely resembles the white butterflies (Pieris) so common in our gardens; but we have no evidence that this resemblance is beneficial. On the contrary, as she resembles both sexes of several species of the same genus inhabiting various quarters of the world, it is more probable that she has simply retained to a large extent her primordial colours.

Various facts support the conclusion that with the greater number of brilliantly-coloured Lepidoptera, it is the male which has been modified; the two sexes having come to differ from each other, or to resemble each other, according to which form of inheritance has prevailed. Inheritance is governed by so many un 410 known laws or conditions, that they seem to us to be most capricious in their action; 524 and we can so far understand how it is that with closely-allied species the sexes of some differ to an astonishing degree, whilst the sexes of others are identical in colour. As the successive steps in the process of variation are necessarily all transmitted through the female, a greater or less number of such steps might readily become developed in her; and thus we can understand the frequent gradations from an extreme difference to no difference at all between the sexes of the species within the same group. These cases of gradation are much too common to favour the supposition that we here see females actually undergoing the process of transition and losing their brightness

162

for the sake of protection; for we have every reason to conclude that at any one time the greater number of species are in a fixed condition. With respect to the differences between the females of the species in the same genus or family, we can perceive that they depend, at least in part, on the females partaking of the colours of their respective males. This is well illustrated in those groups in which the males are ornamented to an extraordinary degree; for the females in these groups generally partake to a certain extent of the splendour of their male partners. Lastly, we continually find, as already remarked, that the females of almost all the species in the same genus, or even family, resemble each other much more closely in colour than do the males; and this indicates that the males have undergone a greater amount of modification than the females.

411Mimickry.—This principle was first made clear in an admirable paper by Mr. Bates,525 who thus threw a flood of light on many obscure problems. It had previously been observed that certain butterflies in S. America belonging to quite distinct families, resembled the Heliconidæ so closely in every stripe and shade of colour that they could not be distinguished except by an experienced entomologist. As the Heliconidæ are coloured in their usual manner, whilst the others depart from the usual colouring of the groups to which they belong, it is clear that the latter are the imitators, and the Heliconidæ the imitated. Mr. Bates further observed that the imitating species are comparatively rare, whilst the imitated swarm in large numbers; the two sets living mingled together. From the fact of the Heliconidæ being conspicuous and beautiful insects, yet so numerous in individuals and species, he concluded that they must be protected from the attacks of birds by some secretion or odour; and this hypothesis has now been confirmed by a considerable body of curious evidence.526 From these considerations Mr. Bates inferred that the butterflies which imitate the protected species had acquired their present marvellously deceptive appearance, through variation and natural selection, in order to be mistaken for the protected kinds and thus to escape being devoured. No explanation is here attempted of the brilliant colours of the imitated, but only of the imitating butterflies. We must account for the colours of the former in the same general manner, as in the cases previously discussed in this chapter. Since the publication of Mr. Bates' paper, similar and equally striking facts have been observed 412by Mr. Wallace527 in the Malayan region, and by Mr. Trimen in South Africa.

As some writers528 have felt much difficulty in understanding how the first steps in the process of mimickry could have been effected through natural selection, it may be well to remark that the process probably has never commenced with forms widely dissimilar in colour. But with two species moderately like each other, the closest resemblance if beneficial to either form could readily be thus gained; and if the imitated form was subsequently and gradually modified through sexual selection or any other means, the imitating form would be led along the same track, and thus be modified to almost any extent, so that it might ultimately assume an appearance or colouring wholly unlike that of the other members of the group to which it belonged. As extremely slight variations in colour would not in many cases suffice to render a species so like another protected species as to lead to its preservation, it should be remembered that many species of Lepidoptera are liable to considerable and abrupt variations in colour. A few instances have been given in this chapter; but under this point of view Mr. Bates' original paper on mimickry, as well as Mr. Wallace's papers, should be consulted.

In the foregoing cases both sexes of the imitating species resemble the imitated; but occasionally the 413female alone mocks a brilliantly-coloured and protected species inhabiting the same district. Consequently the female differs in colour from her own male, and, which is a rare and anomalous circumstance, is the more brightly-coloured of the two. In all the few species of Pieridæ, in which the female is more conspicuously coloured than the male, she imitates, as I am informed by Mr. Wallace, some protected species inhabiting the same region. The female of Diadema anomala is rich purple-brown with almost the whole surface glossed with satiny blue, and she closely imitates the Euplœa midamus, "one of the commonest butterflies of the East;" whilst the male is bronzy or olive-brown, with only a slight blue gloss on the outer parts of the wings.529 Both sexes of this Diadema and of D. bolina follow the same habits of life, so that the differences in colour between the sexes cannot be accounted for by exposure to different conditions;530 even if this explanation were admissible in other instances.531

The above cases of female butterflies which are more brightly-coloured than the males, shew us, firstly, that variations have arisen in a state of nature in the female sex, and have been transmitted exclusively, or almost exclusively, to the same sex; and, secondly, that this form of inheritance has not been determined through natural selection. For if we assume that the females, before they became brightly coloured in imitation of some protected kind, were exposed during each season for a longer period to danger than the males; or if we assume that 414they could not escape so swiftly from their enemies, we can understand how they alone might originally have acquired through natural selection and sexually-limited inheritance their present protective colours. But except on the principle of these variations having been transmitted exclusively to the female offspring, we cannot understand why the males should have remained dull-coloured; for it would surely not have been in any way injurious to each individual male to have partaken by inheritance of the protective colours of the female, and thus to have had a better chance of escaping destruction. In a group in which brilliant colours are so common as with butterflies, it cannot be supposed that the males have been kept dull-coloured through sexual selection by the females rejecting the individuals which were rendered as beautiful as themselves. We may, therefore, conclude that in these cases inheritance by one sex is not due to the modification through natural selection of a tendency to equal inheritance by both sexes.

It may be well here to give an analogous case in another Order, of characters acquired only by the female, though not in the least injurious, as far as we can judge, to the male. Amongst the Phasmidæ, or spectre-insects, Mr. Wallace states that "it is often the females alone that so strikingly resemble leaves, while the males show only a rude approximation." Now, whatever may be the habits of these insects, it is highly improbable that it could be disadvantageous to the males to escape detection by resembling leaves.532Hence we may conclude 415that the females alone in this latter as in the previous cases originally varied in certain characters; these characters having been preserved and augmented through ordinary selection for the sake of protection and from the first transmitted to the female offspring alone.

Bright Colours of Caterpillars.—Whilst reflecting on the beauty of many butterflies, it occurred to me that some caterpillars were splendidly coloured, and as sexual selection could not possibly have here acted, it appeared rash to attribute the beauty of the mature insect to this agency, unless the bright colours of their larvæ could be in some manner explained. In the first place it may be observed that the colours of caterpillars do not stand in any close correlation with those of the mature insect. Secondly, their bright colours do not 416serve in any ordinary manner as a protection. As an instance of this, Mr. Bates informs me that the most conspicuous caterpillar which he ever beheld (that of a Sphinx) lived on the large green leaves of a tree on the open llanos of South America; it was about four inches in length, transversely banded with black and yellow, and with its head, legs, and tail of a bright red. Hence it caught the eye of any man who passed by at the distance of many yards, and no doubt of every passing bird.

I then applied to Mr. Wallace, who has an innate genius for solving difficulties. After some consideration he replied: "Most caterpillars require protection, as may be inferred from some kinds being furnished with spines or irritating hairs, and from many being coloured green like the leaves on which they feed, or curiously like the twigs of the trees on which they live." I may add as another instance of protection, that there is a caterpillar of a moth, as I am informed by Mr. J. Mansel Weale, which lives on the mimosas in South Africa, and fabricates for itself a case, quite undistinguishable from the surrounding thorns. From such considerations Mr. Wallace thought it probable that conspicuously-coloured caterpillars were protected by having a nauseous taste; but as their skin is extremely tender, and as their intestines readily protrude from a wound, a slight peck from the beak of a bird would be as fatal to them as if they had been devoured. Hence, as Mr. Wallace remarks, "distastefulness alone would be insufficient to protect a caterpillar unless some outward sign indicated to its would-be destroyer that its prey was a disgusting morsel." Under these circumstances it would be highly advantageous to a caterpillar to be instantaneously and certainly recognised as unpalatable by all birds and other animals.417 Thus the most gaudy colours would be serviceable, and might have been gained by variation and the survival of the most easily-recognised individuals.

This hypothesis appears at first sight very bold; but when it was brought before the Entomological Society533 it was supported by various statements; and Mr. J. Jenner Weir, who keeps a large number of birds in an aviary, has made, as he informs me, numerous trials, and finds no exception to the rule, that all caterpillars of nocturnal and retiring habits with smooth skins, all of a green colour, and all which imitate twigs, are greedily devoured by his birds. The hairy and spinose kinds are invariably rejected, as were four conspicuously-coloured species. When the birds rejected a caterpillar, they plainly shewed, by shaking their heads and cleansing their beaks, that they were disgusted by the taste.534 Three conspicuous kinds of caterpillars and moths were also given by Mr. A. Butler to some lizards and frogs, and were rejected; though other kinds were eagerly eaten. Thus the probable truth of Mr. Wallace's view is confirmed, namely, that certain caterpillars have been made conspicuous for their own good, so as to be easily recognised by their enemies, on nearly the same principle that certain poisons are coloured by druggists for the good of man. This view will, it is probable, be hereafter extended to many animals, which are coloured in a conspicuous manner.

Summary and Concluding Remarks on Insects.—Looking back to the several Orders, we have seen that the sexes often differ in various characters, the meaning 418of which is not understood. The sexes, also, often differ in their organs of sense or locomotion, so that the males may quickly discover or reach the females, and still oftener in the males possessing diversified contrivances for retaining the females when found. But we are not here much concerned with sexual differences of these kinds.

In almost all the Orders, the males of some species, even of weak and delicate kinds, are known to be highly pugnacious; and some few are furnished with special weapons for fighting with their rivals. But the law of battle does not prevail nearly so widely with insects as with the higher animals. Hence probably it is that the males have not often been rendered larger and stronger than the females. On the contrary they are usually smaller, in order that they may be developed within a shorter time, so as to be ready in large numbers for the emergence of the females.

In two families of the Homoptera the males alone possess, in an efficient state, organs which may be called vocal; and in three families of the Orthoptera the males alone possess stridulating organs. In both cases these organs are incessantly used during the breeding-season, not only for calling the females, but for charming or exciting them in rivalry with other males. No one who admits the agency of natural selection, will dispute that these musical instruments have been acquired through sexual selection. In four other Orders the members of one sex, or more commonly of both sexes, are provided with organs for producing various sounds, which apparently serve merely as call-notes. Even when both sexes are thus provided, the individuals which were able to make the loudest or most continuous noise would gain partners before those which were less noisy, so that their organs have probably been gained419 through sexual selection. It is instructive to reflect on the wonderful diversity of the means for producing sound, possessed by the males alone or by both sexes in no less than six Orders, and which were possessed by at least one insect at an extremely remote geological epoch. We thus learn how effectual sexual selection has been in leading to modifications of structure, which sometimes, as with the Homoptera, are of an important nature.

From the reasons assigned in the last chapter, it is probable that the great horns of the males of many lamellicorn, and some other beetles, have been acquired as ornaments. So perhaps it may be with certain other peculiarities confined to the male sex. From the small size of insects, we are apt to undervalue their appearance. If we could imagine a male Chalcosoma (fig. 15) with its polished, bronzed coat of mail, and vast complex horns, magnified to the size of a horse or even of a dog, it would be one of the most imposing animals in the world.

The colouring of insects is a complex and obscure subject. When the male differs slightly from the female, and neither are brilliantly coloured, it is probable that the two sexes have varied in a slightly different manner, with the variations transmitted to the same sex, without any benefit having been thus derived or evil suffered. When the male is brilliantly coloured and differs conspicuously from the female, as with some dragon-flies and many butterflies, it is probable that he alone has been modified, and that he owes his colours to sexual selection; whilst the female has retained a primordial or very ancient type of colouring, slightly modified by the agencies before explained, and has therefore not been rendered obscure, at least in most cases, for the sake of protection. But the female alone has some420times been coloured brilliantly so as to imitate other protected species inhabiting the same district. When the sexes resemble each other and both are obscurely coloured, there is no doubt that they have been in a multitude of cases coloured for the sake of protection. So it is in some instances when both are brightly coloured, causing them to resemble surrounding objects such as flowers, or other protected species, or indirectly by giving notice to their enemies that they are of an unpalatable nature. In many other cases in which the sexes resemble each other and are brilliantly coloured, especially when the colours are arranged for display, we may conclude that they have been gained by the male sex as an attraction, and have been transferred to both sexes. We are more especially led to this conclusion whenever the same type of coloration prevails throughout a group, and we find that the males of some species differ widely in colour from the females, whilst both sexes of other species are quite alike, with intermediate gradations connecting these extreme states.

In the same manner as bright colours have often been partially transferred from the males to the females, so it has been with the extraordinary horns of many lamellicorn and some other beetles. So, again, the vocal or instrumental organs proper to the males of the Homoptera and Orthoptera have generally been transferred in a rudimentary, or even in a nearly perfect condition to the females; yet not sufficiently perfect to be used for producing sound. It is also an interesting fact, as bearing on sexual selection, that the stridulating organs of certain male Orthoptera are not fully developed until the last moult; and that the colours of certain male dragon-flies are not fully developed until some little time after their emergence from the pupal state, and when they are ready to breed.

421Sexual selection implies that the more attractive individuals are preferred by the opposite sex; and as with insects, when the sexes differ, it is the male which, with rare exceptions, is the most ornamented and departs most from the type to which the species belongs;—and as it is the male which searches eagerly for the female, we must suppose that the females habitually or occasionally prefer the more beautiful males, and that these have thus acquired their beauty. That in most or all the orders the females have the power of rejecting any particular male, we may safely infer from the many singular contrivances possessed by the males, such as great jaws, adhesive cushions, spines, elongated legs, &c., for seizing the female; for these contrivances shew that there is some difficulty in the act. In the case of unions between distinct species, of which many instances have been recorded, the female must have been a consenting party. Judging from what we know of the perceptive powers and affections of various insects, there is no antecedent improbability in sexual selection having come largely into action; but we have as yet no direct evidence on this head, and some facts are opposed to the belief. Nevertheless, when we see many males pursuing the same female, we can hardly believe that the pairing is left to blind chance—that the female exerts no choice, and is not influenced by the gorgeous colours or other ornaments, with which the male alone is decorated.

If we admit that the females of the Homoptera and Orthoptera appreciate the musical tones emitted by their male partners, and that the various instruments for this purpose have been perfected through sexual selection, there is little improbability in the females of other insects appreciating beauty in form or colour, and consequently in such characters having been thus gained422 by the males. But from the circumstance of colour being so variable, and from its having been so often modified for the sake of protection, it is extremely difficult to decide in how large a proportion of cases sexual selection has come into play. This is more especially difficult in those Orders, such as the Orthoptera, Hymenoptera, and Coleoptera, in which the two sexes rarely differ much in colour; for we are thus cut off from our best evidence of some relation between the reproduction of the species and colour. With the Coleoptera, however, as before remarked, it is in the great lamellicorn group, placed by some authors at the head of the Order, and in which we sometimes see a mutual attachment between the sexes, that we find the males of some species possessing weapons for sexual strife, others furnished with wonderful horns, many with stridulating organs, and others ornamented with splendid metallic tints. Hence it seems probable that all these characters have been gained through the same means, namely sexual selection.

When we treat of Birds, we shall see that they present in their secondary sexual characters the closest analogy with insects. Thus, many male birds are highly pugnacious, and some are furnished with special weapons for fighting with their rivals. They possess organs which are used during the breeding-season for producing vocal and instrumental music. They are frequently ornamented with combs, horns, wattles and plumes of the most diversified kinds, and are decorated with beautiful colours, all evidently for the sake of display. We shall find that, as with insects, both sexes, in certain groups, are equally beautiful, and are equally provided with ornaments which are usually confined to the male sex. In other groups both sexes are equally plain-coloured and unornamented. Lastly, in some few423 anomalous cases, the females are more beautiful than the males. We shall often find, in the same group of birds, every gradation from no difference between the sexes, to an extreme difference. In the latter case we shall see that the females, like female insects, often possess more or less plain traces of the characters which properly belong to the males. The analogy, indeed, in all these respects between birds and insects, is curiously close. Whatever explanation applies to the one class probably applies to the other; and this explanation, as we shall hereafter attempt to shew, is almost certainly sexual selection.

FOOTNOTES:

1As the works of the first-named authors are so well known, I need not give the titles; but as those of the latter are less well known in England, I will give them:—'Sechs Vorlesungen über die Darwin'sche Theorie:' zweite Auflage, 1868, von Dr. L. Büchner; translated into French under the title 'Conférences sur la Théorie Darwinienne,' 1869. 'Der Mensch, im Lichte der Darwin'sche Lehre,' 1865, von Dr. F. Rolle. I will not attempt to give references to all the authors who have taken the same side of the question. Thus G. Canestrini has published ('Annuario della Soc. d. Nat.,' Modena, 1867, p. 81) a very curious paper on rudimentary characters, as bearing on the origin of man. Another work has (1869) been published by Dr. Barrago Francesco, bearing in Italian the title of "Man, made in the image of God, was also made in the image of the ape."

2Prof. Häckel is the sole author who, since the publication of the 'Origin,' has discussed, in his various works, in a very able manner, the subject of sexual selection, and has seen its full importance.

3'Grosshirnwindungen des Menschen,' 1868, s. 96.

4'Leç. sur la Phys.' 1866, p. 890, as quoted by M. Dally, 'L'Ordre des Primates et le Transformisme,' 1868, p. 29.

5'Naturgeschichte der Säugethiere von Paraguay,' 1830, s. 50.

6Brehm, 'Thierleben,' B. i. 1864, s. 75, 86. On the Ateles, s. 105. For other analogous statements, see s. 25, 107.

7With respect to insects see Dr. Laycock 'On a General Law of Vital Periodicity,' British Association, 1842. Dr. Macculloch, 'Silliman's North American Journal of Science,' vol. xvii. p. 305, has seen a dog suffering from tertian ague.

8I have given the evidence on this head in my 'Variation of Animals and Plants under Domestication,' vol. ii. p. 15.

9"Mares e diversis generibus Quadrumanorum sine dubio dignoscunt feminas humanas a maribus. Primum, credo, odoratu, postea aspectu. Mr. Youatt, qui diu in Hortis Zoologicis (Bestiariis) medicus animalium erat, vir in rebus observandis cautus et sagax, hoc mihi certissime probavit, et curatores ejusdem loci et alii e ministris confirmaverunt. Sir Andrew Smith et Brehm notabant idem in Cynocephalo. Illustrissimus Cuvier etiam narrat multa de hac re quâ ut opinor nihil turpius potest indicari inter omnia hominibus et Quadrumanis communia. Narrat enim Cynocephalum quendam in furorem incidere aspectu feminarum aliquarum, sed nequaquam accendi tanto furore ab omnibus. Semper eligebat juniores, et dignoscebat in turba, et advocabat voce gestuque."

10This remark is made with respect to Cynocephalus and the anthropomorphous apes by Geoffroy Saint-Hilaire and F. Cuvier, 'Hist. Nat. des Mammifères,' tom. i. 1824.

11Huxley, 'Man's Place in Nature,' 1863, p. 34.

12'Man's Place in Nature,' 1863, p. 67.

13The human embryo (upper fig.) is from Ecker, 'Icones Phys.,' 1851-1859, tab. xxx. fig. 2. This embryo was ten lines in length, so that the drawing is much magnified. The embryo of the dog is from Bischoff, 'Entwicklungsgeschichte des Hunde-Eies,' 1845, tab. xi. fig. 42 B. This drawing is five times magnified, the embryo being 25 days old. The internal viscera have been omitted, and the uterine appendages in both drawings removed. I was directed to these figures by Prof. Huxley, from whose work, 'Man's Place in Nature.' the idea of giving them was taken. Häckel has also given analogous drawings in his 'Schöpfungsgeschichte.'

14Prof. Wyman in 'Proc. of American Acad. of Sciences,' vol. iv. 1860, p. 17.

15Owen, 'Anatomy of Vertebrates,' vol. i. p. 533.

16'Die Grosshirnwindungen des Menschen,' 1868, s. 95.

17'Anatomy of Vertebrates,' vol. ii. p. 553.

18'Proc. Soc. Nat. Hist.' Boston, 1863, vol. ix. p. 185.

19'Man's Place in Nature,' p. 65.

20I had written a rough copy of this chapter before reading a valuable paper, "Caratteri rudimentali in ordine all'origine del uomo" ('Annuario della Soc. d. Nat.,' Modena, 1867, p. 81), by G. Canestrini, to which paper I am considerably indebted. Häckel has given admirable discussions on this whole subject, under the title of Dysteleology, in his 'Generelle Morphologie' and 'Schöpfungsgeschichte.'

21Some good criticisms on this subject have been given by Messrs. Murie and Mivart, in 'Transact. Zoolog. Soc.' 1869, vol. vii. p. 92.

22'Variation of Animals and Plants under Domestication,' vol. ii. pp. 317 and 397. See also 'Origin of Species,' 5th edit. p. 535.

23For instance M. Richard ('Annales des Sciences Nat.' 3rd series, Zoolog. 1852, tom. xviii. p. 13) describes and figures rudiments of what he calls the "muscle pédieux de la main," which he says is sometimes "infiniment petit." Another muscle, called "le tibial postérieur," is generally quite absent in the hand, but appears from time to time in a more or less rudimentary condition.

24Prof. W. Turner, 'Proc. Royal Soc. Edinburgh,' 1866-67, p. 65.

25Canestrini quotes Hyrt. ('Annuario della Soc. dei Naturalisti,' Modena, 1867, p. 97) to the same effect.

26'The Diseases of the Ear,' by J. Toynbee, F.R.S., 1860, p. 12.

27See also some remarks, and the drawings of the ears of the Lemuroidea, in Messrs. Murie and Mivart's excellent paper in 'Transact. Zoolog. Soc.' vol. vii. 1869, pp. 6 and 90.

28Müller's 'Elements of Physiology,' Eng. translat., 1842, vol. ii. p. 1117. Owen, 'Anatomy of Vertebrates,' vol. iii. p. 260; ibid. on the Walrus, 'Proc. Zoolog. Soc.' November 8th, 1854. See also R. Knox, 'Great Artists and Anatomists,' p. 106. This rudiment apparently is somewhat larger in Negroes and Australians than in Europeans, see Carl Vogt, 'Lectures on Man,' Eng. translat. p. 129.

29'The Physiology and Pathology of Mind,' 2nd edit. 1868, p. 134.

30Eschricht, Ueber die Richtung der Haare am menschlichen Körper 'Müllers Archiv für Anat. und Phys.' 1837, s. 47. I shall often have to refer to this very curious paper.

31Paget, 'Lectures on Surgical Pathology,' 1853, vol. i. p. 71.

32Eschricht, ibid. s. 40, 47.

33Dr. Webb, 'Teeth in Man and the Anthropoid Apes,' as quoted by Dr. C. Carter Blake in 'Anthropological Review,' July, 1867, p. 299.

34Owen, 'Anatomy of Vertebrates,' vol. iii. pp. 320, 321, and 325.

35'On the Primitive Form of the Skull,' Eng. translat. in 'Anthropological Review,' Oct. 1868, p. 426.

36Owen, 'Anatomy of Vertebrates,' vol. iii. pp. 416, 434, 441.

37'Annuario della Soc. d. Nat.' Modena, 1867, p. 94.

38M. C. Martins ("De l'Unité Organique," in 'Revue des Deux Mondes,' June 15, 1862, p. 16), and Häckel ('Generelle Morphologie,' B. ii. s. 278), have both remarked on the singular fact of this rudiment sometimes causing death.

39'The Lancet,' Jan. 24, 1863, p. 83. Dr. Knox, 'Great Artists and Anatomists,' p. 63. See also an important memoir on this process by Dr. Grube, in the 'Bulletin de l'Acad. Imp. de St. Pétersbourg,' tom. xii. 1867, p. 448.

40"On the Caves of Gibraltar," 'Transact. Internat. Congress of Prehist. Arch.' Third Session, 1869, p. 54.

41Quatrefages has lately collected the evidence on this subject. 'Revue des Cours Scientifiques,' 1867-1868, p. 625.

42Owen, 'On the Nature of Limbs,' 1849, p. 114.

43Leuckart, in Todd's 'Cyclop. of Anat.' 1849-52, vol. iv. p. 1415. In man this organ is only from three to six lines in length, but, like so many other rudimentary parts, it is variable in development as well as in other characters.

44See, on this subject, Owen, 'Anatomy of Vertebrates,' vol. iii. pp. 675, 676, 706.

45See the evidence on these points, as given by Lubbock, 'Prehistoric Times,' p. 354, &c.

46'L'Instinct chez les Insectes.' 'Revue des Deux Mondes,' Feb. 1870, p. 690.

47'The American Beaver and his Works,' 1868.

48'The Principles of Psychology,' 2nd edit. 1870, pp. 418-443.

49'Contributions to the Theory of Natural Selection,' 1870, p. 212

50'Recherches sur les Mœurs des Fourmis,' 1810, p. 173.

51All the following statements, given on the authority of these two naturalists, are taken from Rengger's 'Naturges. der Säugethiere von Paraguay,' 1830, s. 41-57, and from Brehm's 'Thierleben,' B. i. s. 10-87.

52'Bridgewater Treatise,' p. 263.

53W. C. L. Martin, 'Nat. Hist. of Mammalia,' 1841, p. 405.

54Quoted by Vogt, 'Mémoire sur les Microcéphales,' 1867, p. 168.

55'The Variation of Animals and Plants under Domestication,' vol. i. p. 27.

56'Les Mœurs des Fourmis,' 1810, p. 150.

57Quoted in Dr. Maudsley's 'Physiology and Pathology of Mind,' 1868, pp. 19, 220.

58Dr. Jerdon, 'Birds of India,' vol. i. 1862, p. xxi.

59Mr. L. H. Morgan's work on 'The American Beaver,' 1868, offers a good illustration of this remark. I cannot, however, avoid thinking that he goes too far in underrating the power of Instinct.

60'The Moor and the Loch,' p. 45. Col. Hutchinson on 'Dog Breaking,' 1850, p. 46.

61'Personal Narrative,' Eng. translat., vol. iii. p. 106.

62Quoted by Sir C. Lyell, 'Antiquity of Man,' p. 497.

63'Journal of Researches during the Voyage of the "Beagle,"' 1845, p. 398. 'Origin of Species,' 5th edit. p. 260.

64'Lettres Phil. sur l'Intelligence des Animaux,' nouvelle edit. 1802, p. 86.

65See the evidence on this head in chap. i. vol. i. 'On the Variation of Animals and Plants under Domestication.'

66'Proc. Zoolog. Soc.' 1864, p. 186.

67Savage and Wyman in 'Boston Journal of Nat. Hist.' vol. iv. 1843-44, p. 383.

68'Säugethiere von Paraguay,' 1830, s. 51-56.

69'Thierleben,' B. i. s. 79, 82.

70'The Malay Archipelago,' vol. i. 1869, p. 87.

71'Primeval Man,' 1869, pp. 145, 147.

72'Prehistoric Times,' 1865, p. 473, &c.

73Quoted in 'Anthropological Review,' 1864, p. 158.

74Rengger, ibid. s. 45.

75See my 'Variation of Animals and Plants under Domestication,' vol. i. p. 27.

76See a discussion on this subject in Mr. E. B. Tylor's very interesting work, 'Researches into the Early History of Mankind,' 1865, chaps. ii. to iv.

77Hon. Daines Barrington in 'Philosoph. Transactions,' 1773, p. 262. See also Dureau de la Malle, in 'Ann. des Sc. Nat.' 3rd series, Zoolog. tom. x. p. 119.

78'On the Origin of Language,' by H. Wedgwood, 1866. 'Chapters on Language,' by the Rev. F. W. Farrar, 1865. These works are most interesting. See also 'De la Phys. et de Parole,' par Albert Lemoine, 1865, p. 190. The work on this subject, by the late Prof. Aug. Schleicher, has been translated by Dr. Bikkers into English, under the title of 'Darwinism tested by the Science of Language,' 1869.

79Vogt, 'Mémoire sur les Microcéphales,' 1867, p. 169. With respect to savages, I have given some facts in my 'Journal of Researches,' &c., 1845, p. 206.

80See clear evidence on this head in the two works so often quoted, by Brehm and Rengger.

81See remarks on this head by Dr. Maudsley, 'The Physiology and Pathology of Mind,' 2nd edit. 1868, p. 199.

82Many curious cases have been recorded. See, for instance, 'Inquiries Concerning the Intellectual Powers,' by Dr. Abercrombie, 1838, p. 150.

83'The Variation of Animals and Plants under Domestication,' vol. ii. p. 6.

84See some good remarks to this effect by Dr. Maudsley, 'The Physiology and Pathology of Mind,' 1808, p. 199.

85Macgillivray, 'Hist. of British Birds,' vol. ii. 1839, p. 29. An excellent observer, Mr. Blackwall, remarks that the magpie learns to pronounce single words, and even short sentences, more readily than almost any other British bird; yet, as he adds, after long and closely investigating its habits, he has never known it, in a state of nature, display any unusual capacity for imitation. 'Researches in Zoology,' 1834, p. 158.

86See the very interesting parallelism between the development of speech and languages, given by Sir C. Lyell in 'The Geolog. Evidences of the Antiquity of Man,' 1863, chap. xxiii.

87See remarks to this effect by the Rev. F. W. Farrar, in an interesting article, entitled "Philology and Darwinism" in 'Nature,' March 24th, 1870, p. 528.

88'Nature,' Jan. 6th, 1870, p. 257.

89Quoted by C. S. Wake, 'Chapters on Man,' 1868, p. 101.

90Buckland, 'Bridgewater Treatise,' p. 411.

91See some good remarks on the simplification of languages, by Sir J. Lubbock, 'Origin of Civilisation,' 1870, p. 278.

92'Conférences sur la Théorie Darwinienne,' French translat., 1869, p. 132.

93The Rev. Dr. J. M'Cann, 'Anti-Darwinism,' 1869, p. 13.

94'The Spectator,' Dec. 4th, 1869, p. 1430.

95See an excellent article on this subject by the Rev. F. W. Farrar, in the 'Anthropological Review,' Aug. 1864, p. ccxvii. For further facts see Sir J. Lubbock, 'Prehistoric Times,' 2nd edit. 1869. p. 564; and especially the chapters on Religion in his 'Origin of Civilisation,' 1870.

96The Worship of Animals and Plants, in the 'Fortnightly Review,' Oct. 1, 1869, p. 422.

97Tylor, 'Early History of Mankind,' 1865, p. 6. See also the three striking chapters on the Development of Religion, in Lubbock's 'Origin of Civilisation,' 1870. In a like manner Mr. Herbert Spencer, in his ingenious essay in the 'Fortnightly Review' (May 1st, 1870, p. 535), accounts for the earliest forms of religious belief throughout the world, by man being led through dreams, shadows, and other causes, to look at himself as a double essence, corporeal and spiritual. As the spiritual being is supposed to exist after death and to be powerful, it is propitiated by various gifts and ceremonies, and its aid invoked. He then further shews that names or nicknames given from some animal or other object to the early progenitors or founders of a tribe, are supposed after a long interval to represent the real progenitor of the tribe; and such animal or object is then naturally believed still to exist as a spirit, is held sacred, and worshipped as a god. Nevertheless I cannot but suspect that there is a still earlier and ruder stage, when anything which manifests power or movement is thought to be endowed with some form of life, and with mental faculties analogous to our own.

98See an able article on the Psychical Elements of Religion, by Mr. L. Owen Pike, in 'Anthropolog. Review,' April, 1870, p. lxiii.

99'Religion, Moral, &c., der Darwin'schen Art-Lehre,' 1869, s. 53.

100'Prehistoric Times,' 2nd edit. p. 571. In this work (at p. 553) there will be found an excellent account of the many strange and capricious customs of savages.

101See, for instance, on this subject, Quatrefages, 'Unité de l'Espèce Humaine,' 1861, p. 21, &c.

102'Dissertation on Ethical Philosophy,' 1837, p. 231, &c.

103'Metaphysics of Ethics,' translated by J. W. Semple, Edinburgh, 1836, p. 136.

104Mr. Bain gives a list ('Mental and Moral Science,' 1868, p. 543-725) of twenty-six British authors who have written on this subject, and whose names are familiar to every reader; to these, Mr. Bain's own name, and those of Mr. Lecky, Mr. Shadworth Hodgson, and Sir J. Lubbock, as well as of others, may be added.

105Sir B. Brodie, after observing that man is a social animal ('Psychological Enquiries,' 1854, p. 192), asks the pregnant question, "ought not this to settle the disputed question as to the existence of a moral sense?" Similar ideas have probably occurred to many persons, as they did long ago to Marcus Aurelius. Mr. J. S. Mill speaks, in his celebrated work, 'Utilitarianism,' (1864, p. 46), of the social feelings as a "powerful natural sentiment," and as "the natural basis of sentiment for utilitarian morality;" but on the previous page he says, "if, as is my own belief, the moral feelings are not innate, but acquired, they are not for that reason less natural." It is with hesitation that I venture to differ from so profound a thinker, but it can hardly be disputed that the social feelings are instinctive or innate in the lower animals; and why should they not be so in man? Mr. Bain (see, for instance, 'The Emotions and the Will,' 1865, p. 481) and others believe that the moral sense is acquired by each individual during his lifetime. On the general theory of evolution this is at least extremely improbable.

106'Die Darwin'sche Theorie,' s. 101.

107Mr. R. Browne in 'Proc. Zoolog. Soc.' 1868, p. 409.

108Brehm, 'Thierleben,' B. i. 1864, s. 52, 79. For the case of the monkeys extracting thorns from each other, see s. 54. With respect to the Hamadryas turning over stones, the fact is given (s. 76) on the evidence of Alvarez, whose observations Brehm thinks quite trustworthy. For the cases of the old male baboons attacking the dogs, see s. 79; and with respect to the eagle, s. 56.

109'Annals and Mag. of Nat. Hist.' November, 1868, p. 382.

110Sir J. Lubbock, 'Prehistoric Times,' 2nd edit. p. 446.

111As quoted by Mr. L. H. Morgan, 'The American Beaver,' 1868, p. 272. Capt. Stansbury also gives an interesting account of the manner in which a very young pelican, carried away by a strong stream, was guided and encouraged in its attempts to reach the shore by half a dozen old birds.

112As Mr. Bain states, "effective aid to a sufferer springs from sympathy proper:" 'Mental and Moral Science,' 1868, p. 245.

113'Thierleben,' B. i. s. 85.

114'De l'Espèce et de la Class.' 1869, p. 97.

115'Der Darwin'schen Art-Lehre,' 1869, s. 54.

116Brehm, 'Thierleben,' B. i. s. 76.

117See the first and striking chapter in Adam Smith's 'Theory of Moral Sentiments.' Also Mr. Bain's 'Mental and Moral Science,' 1868, p. 244, and 275-282. Mr. Bain states, that "sympathy is, indirectly, a source of pleasure to the sympathiser;" and he accounts for this through reciprocity. He remarks that "the person benefited, or others in his stead, may make up, by sympathy and good offices returned, for all the sacrifice." But if, as appears to be the case, sympathy is strictly an instinct, its exercise would give direct pleasure, in the same manner as the exercise, as before remarked, of almost every other instinct.

118This fact, the Rev. L. Jenyns states (see his edition of 'White's Nat. Hist. of Selborne,' 1853, p. 204) was first recorded by the illustrious Jenner, in 'Phil. Transact.' 1824, and has since been confirmed by several observers, especially by Mr. Blackwall. This latter careful observer examined, late in the autumn, during two years, thirty-six nests; he found that twelve contained young dead birds, five contained eggs on the point of being hatched, and three eggs not nearly hatched. Many birds not yet old enough for a prolonged flight are likewise deserted and left behind. See Blackwall, 'Researches in Zoology,' 1834, pp. 108, 118. For some additional evidence, although this is not wanted, see Leroy, 'Lettres Phil.' 1802, p. 217.

119Hume remarks ('An Enquiry Concerning the Principles of Morals,' edit. of 1751, p. 132), "there seems a necessity for confessing that the happiness and misery of others are not spectacles altogether indifferent to us, but that the view of the former ... communicates a secret joy; the appearance of the latter ... throws a melancholy damp over the imagination."

120'Mental and Moral Science,' 1868, p. 254.

121I have given one such case, namely of three Patagonian Indians who preferred being shot, one after the other, to betraying the plans of their companions in war ('Journal of Researches,' 1845, p. 103).

122Dr. Prosper Despine, in his 'Psychologie Naturelle,' 1868 (tom. i. p. 243; tom ii. p. 169) gives many curious cases of the worst criminals, who apparently have been entirely destitute of conscience.

123See an able article in the 'North British Review,' 1867, p. 395. See also Mr. W. Bagehot's articles on the Importance of Obedience and Coherence to Primitive Man, in the 'Fortnightly Review,' 1867, p. 529, and 1868, p. 457, &c.

124The fullest account which I have met with is by Dr. Gerland, in his 'Ueber das Aussterben der Naturvölker,' 1868; but I shall have to recur to the subject of infanticide in a future chapter.

125See the very interesting discussion on Suicide in Lecky's 'History of European Morals,' vol. i. 1869, p. 223.

126See, for instance, Mr. Hamilton's account of the Kaffirs, 'Anthropological Review,' 1870, p. xv.

127Mr. M'Lennan has given 'Primitive Marriage,' 1865, p. 176, a good collection of facts on this head.

128Lecky, 'History of European Morals,' vol. i. 1869, p. 109.

129'Embassy to China,' vol. ii. p. 348.

130See on this subject copious evidence in Chap. vii. of Sir J. Lubbock, 'Origin of Civilisation,' 1870.

131For instance Lecky, 'Hist. European Morals,' vol. i. p. 124.

132This term is used in an able article in the 'Westminster Review,' Oct. 1869, p. 498. For the Greatest Happiness principle, see J. S. Mill,

Good instances are given by Mr. Wallace in 'Scientific Opinion,' Sept. 15, 1869; and more fully in his 'Contributions to the Theory of Natural Selection,' 1870, p. 353.

134Tennyson, 'Idylls of the King,' p. 244.

135'The Thoughts of the Emperor M. Aurelius Antoninus,' Eng. translat., 2nd edit., 1869, p. 112. Marcus Aurelius was born A.D. 121.

136Letter to Mr. Mill in Bain's 'Mental and Moral Science,' 1868, p. 722.

137A writer in the 'North British Review' (July, 1869, p. 531), well capable of forming a sound judgment, expresses himself strongly to this effect. Mr. Lecky ('Hist. of Morals,' vol. i. p. 143) seems to a certain extent to coincide.

138See his remarkable work on 'Hereditary Genius,' 1869, p. 349. The Duke of Argyll ('Primeval Man,' 1869, p. 188) has some good remarks on the contest in man's nature between right and wrong.

139'The Thoughts of Marcus Aurelius,' &c., p. 139.

140'Investigations in Military and Anthropolog. Statistics of American Soldiers,' by B. A. Gould, 1869, p. 256.

141With respect to the "Cranial forms of the American aborigines," see Dr. Aitken Meigs in 'Proc. Acad. Nat. Sci.' Philadelphia, May, 1866. On the Australians, see Huxley, in Lyell's 'Antiquity of Man,' 1863, p. 87. On the Sandwich Islanders, Prof. J. Wyman, 'Observations on Crania,' Boston, 1868, p. 18.

142'Anatomy of the Arteries,' by R. Quain.

143'Transact. Royal Soc.' Edinburgh, vol. xxiv. p. 175, 189.

144'Proc. Royal Soc.' 1867, p. 544; also 1868, p. 483, 524. There is a previous paper, 1866, p. 229.

145'Proc. R. Irish Academy,' vol. x. 1868, p. 141.

146'Act. Acad.,' St. Petersburg, 1778, part ii. p. 217.

147Brehm, 'Thierleben,' B. i. s. 58, 87. Rengger, 'Säugethiere von Paraguay,' s. 57.

148'Variation of Animals and Plants under Domestication,' vol. ii. chap. xii.

149'Hereditary Genius: an Inquiry into its Laws and Consequences,' 1869.

150Mr. Bates remarks ('The Naturalist on the Amazons,' 1863, vol. ii. p. 159), with respect to the Indians of the same S. American tribe, "no two of them were at all similar in the shape of the head; one man had an oval visage with fine features, and another was quite Mongolian in breadth and prominence of cheek, spread of nostrils, and obliquity of eyes."

151Blumenbach, 'Treatises on Anthropolog.' Eng. translat., 1865, p. 205.

152Godron, 'De l'Espèce,' 1859, tom. ii. livre 3. Quatrefages, 'Unité de l'Espèce Humaine,' 1861. Also Lectures on Anthropology, given in the 'Revue des Cours Scientifiques,' 1866-1868.

153'Hist. Gen. et Part. des Anomalies de l'Organisation,' in three volumes, tom. i. 1832.

154I have fully discussed these laws in my 'Variation of Animals and Plants under Domestication,' vol. ii. chap. xxii. and xxiii. M. J. P. Durand has lately 1868; published a valuable essay 'De l'Influence des Milieux, &c.' He lays much stress on the nature of the soil.

155'Investigations in Military and Anthrop. Statistics,' &c. 1869, by B. A. Gould, p. 93, 107, 126, 131, 134.

156For the Polynesians, see Prichard's 'Physical Hist. of Mankind,' vol. v. 1847, p. 145, 283. Also Godron, 'De l'Espèce,' tom. ii. p. 289. There is also a remarkable difference in appearance between the closely-allied Hindoos inhabiting the Upper Ganges and Bengal; see Elphinstone's 'History of India,' vol. i. p. 324.

157'Memoirs, Anthropolog. Soc.' vol. iii. 1867-69, p. 561, 565, 567.

158Dr. Brakenridge, 'Theory of Diathesis,' 'Medical Times,' June 19 and July 17, 1869.

159I have given authorities for these several statements in my 'Variation of Animals under Domestication,' vol. ii. p. 297-300. Dr. Jaeger, "Ueber das Längenwachsthum der Knochen," 'Jenaischen Zeitschrift,' B. v. Heft i.

160'Investigations,' &c. By B. A. Gould, 1869, p. 288.

161'Säugethiere von Paraguay,' 1830, s. 4.

162'History of Greenland,' Eng. translat. 1767, vol. i. p. 230.

163'Intermarriage.' By Alex. Walker, 1838. p. 377.

164'The Variation of Animals under Domestication,' vol. i. p. 173.

165'Principles of Biology,' vol. i. p. 455.

166Paget, 'Lectures on Surgical Pathology,' vol. i. 1853, p. 209.

167'The Variation of Animals under Domestication,' vol. i. p. 8.

168'Säugethiere von Paraguay,' s. 8, 10. I have had good opportunities for observing the extraordinary power of eyesight in the Fuegians.' See also Lawrence ('Lectures on Physiology,' &c., 1822, p. 404) on this same subject. M. Giraud-Teulon has recently collected ('Revue des Cours Scientifiques,' 1870, p. 625) a large and valuable body of evidence proving that the cause of short-sight, "C'est le travail assidu, de près."

169Prichard, 'Phys. Hist. of Mankind,' on the authority of Blumenbach, vol. i. 1851, p. 311; for the statement by Pallas, vol. iv. 1844, p. 407.

170Quoted by Prichard, 'Researches into the Phys. Hist. of Mankind,' vol. v. p. 463.

171Mr. Forbes' valuable paper is now published in the 'Journal of the Ethnological Soc. of London,' new series, vol. ii. 1870, p. 193.

172Dr. Wilckens ('Landwirthschaft. Wochenblatt,' No. 10, 1869) has lately published an interesting essay shewing how domestic animals, which live in mountainous regions, have their frames modified.

173'Mémoire sur les Microcéphales,' 1867, p. 50, 125, 169, 171, 184-198.

174See Dr. A. Farre's well-known article in the 'Cyclop. of Anat. and Phys.' vol. v. 1859, p. 642. Owen 'Anatomy of Vertebrates,' vol. iii. 1868, p. 687. Prof. Turner in 'Edinburgh Medical Journal,' Feb. 1865.

175'Annuario della Soc. dei Naturalisti in Modena,' 1867, p. 83. Prof. Canestrini gives extracts on this subject from various authorities. Laurillard remarks, that as he has found a complete similarity in the form, proportions, and connexion of the two malar bones in several human subjects and in certain apes, he cannot consider this disposition of the parts as simply accidental.

176A whole series of cases is given by Isid. Geoffroy St.-Hilaire, 'Hist. des Anomalies,' tom. iii. p. 437.

177In my 'Variation of Animals under Domestication' (vol. ii. p. 57) I attributed the not very rare cases of supernumerary mammæ in women to reversion. I was led to this as a probable conclusion, by the additional mammæ being generally placed symmetrically on the breast, and more especially from one case, in which a single efficient mamma occurred in the inguinal region of a woman, the daughter of another woman with supernumerary mammæ. But Prof. Preyer ('Der Kampf um das Dasein,' 1869, s. 45) states that mammæ erraticæ have been known to occur in other situations, even on the back; so that the force of my argument is greatly weakened or perhaps quite destroyed.

With much hesitation I, in the same work (vol. ii. p. 12), attributed the frequent cases of polydactylism in men to reversion. I was partly led to this through Prof. Owen's statement, that some of the Ichthyopterygia possess more than five digits, and therefore, as I supposed, had retained a primordial condition; but after reading Prof. Gegenbaur's paper ('Jenaischen Zeitschrift,' B. v. Heft 3, s. 341), who is the highest authority in Europe on such a point, and who disputes Owen's conclusion, I see that it is extremely doubtful whether supernumerary digits can thus be accounted for. It was the fact that such digits not only frequently occur and are strongly inherited, but have the power of regrowth after amputation, like the normal digits of the lower vertebrata, that chiefly led me to the above conclusion. This extraordinary fact of their regrowth remains inexplicable, if the belief in reversion to some extremely remote progenitor must be rejected. I cannot, however, follow Prof. Gegenbaur in supposing that additional digits could not reappear through reversion, without at the same time other parts of the skeleton being simultaneously and similarly modified; for single characters often reappear through reversion.

178'Anatomy of Vertebrates,' vol. iii. 1868, p. 323.

179'Generelle Morphologie,' 1866, B. ii. s. clv.

180Carl Vogt's 'Lectures on Man,' Eng. translat. 1864, p. 151.

181C. Carter Blake, on a jaw from La Naulette, 'Anthropolog. Review,' 1867, p. 295. Schaaffhausen, ibid. 1868, p. 426.

182'The Anatomy of Expression,' 1844, p. 110, 131.

183Quoted by Prof. Canestrini in the 'Annuario,' &c., 1867, p. 90.

184These papers deserve careful study by any one who desires to learn how frequently our muscles vary, and in varying come to resemble those of the Quadrumana. The following references relate to the few points touched on in my text: vol. xiv. 1865, p. 379-384; vol. xv. 1866, p. 241, 242; vol. xv. 1867, p. 544; vol. xvi. 1868, p. 524. I may here add that Dr. Murie and Mr. St. George Mivart have shewn in their Memoir on the Lemuroidea ('Transact. Zoolog. Soc.' vol. vii. 1869, p. 96), how extraordinarily variable some of the muscles are in these animals, the lowest members of the Primates. Gradations, also, in the muscles leading to structures found in animals still lower in the scale, are numerous in the Lemuroidea.

185Prof. Macalister in 'Proc. R. Irish Academy,' vol. x. 1868, p. 124.

186Prof. Macalister (ibid. p. 121) has tabulated his observations, and finds that muscular abnormalities are most frequent in the fore-arms, secondly in the face, thirdly in the foot, &c.

187The Rev. Dr. Haughton, after giving ('Proc. R. Irish Academy,' June 27, 1864, p. 715) a remarkable case of variation in the human flexor pollicis longus, adds, "This remarkable example shews that man may sometimes possess the arrangement of tendons of thumb and fingers characteristic of the macaque; but whether such a case should be regarded as a macaque passing upwards into a man, or a man passing downwards into a macaque, or as a congenital freak of nature, I cannot undertake to say." It is satisfactory to hear so capable an anatomist, and so embittered an opponent of evolutionism, admitting even the possibility of either of his first propositions. Prof. Macalister has also described ('Proc. R. Irish Acad.' vol. x. 1864, p. 138) variations in the flexor pollicis longus, remarkable from their relations to the same muscle in the Quadrumana.

188The authorities for these several statements are given in my 'Variation of Animals under Domestication,' vol. ii. p. 320-335.

189This whole subject has been discussed in chap. xxiii. vol. ii. of my 'Variation of Animals and Plants under Domestication.'

190See the ever memorable 'Essay on the Principle of Population,' by the Rev. T. Malthus, vol. i. 1826, p. 6, 517.

191'Variation of Animals and Plants under Domestication,' vol. ii. p. 111-113, 163.

192Mr. Sedgwick, 'British and Foreign Medico-Chirurg. Review,' July, 1863, p. 170.

193'The Annals of Rural Bengal,' by W. W. Hunter, 1868, p. 259.

194'Primitive Marriage,' 1865.

195See some good remarks to this effect by W. Stanley Jevons, "A Deduction from Darwin's Theory," 'Nature,' 1869, p. 231.

196Latham, 'Man and his Migrations,' 1851, p. 135.

197Messrs. Murie and Mivart in their "Anatomy of the Lemuroidea" ('Transact. Zoolog. Soc.' vol. vii. 1869, p. 96-98) say, "some muscles are so irregular in their distribution that they cannot be well classed in any of the above groups." These muscles differ even on the opposite sides of the same individual.

198'Quarterly Review,' April, 1869, p. 392. This subject is more fully discussed in Mr. Wallace's 'Contributions to the Theory of Natural Selection,' 1870, in which all the essays referred to in this work are republished. The 'Essay on Man' has been ably criticised by Prof. Claparède, one of the most distinguished zoologists in Europe, in an article published in the 'Bibliothèque Universelle,' June, 1870. The remark quoted in my text will surprise every one who has read Mr. Wallace's celebrated paper on 'The Origin of Human Races deduced from the Theory of Natural Selection,' originally published in the 'Anthropological Review,' May, 1864, p. clviii. I cannot here resist quoting a most just remark by Sir J. Lubbock ('Prehistoric Times,' 1865, p. 479) in reference to this paper, namely, that Mr. Wallace, "with characteristic unselfishness, ascribes it (i.e. the idea of natural selection) unreservedly to Mr. Darwin, although, as is well known, he struck out the idea independently, and published it, though not with the same elaboration, at the same time."

199Quoted by Mr. Lawson Tait in his "Law of Natural Selection,"—'Dublin Quarterly Journal of Medical Science,' Feb. 1869. Dr. Keller is likewise quoted to the same effect.

200Owen, 'Anatomy of Vertebrates,' vol. iii. p. 71.

201'Quarterly Review,' April, 1869, p. 392.

202In Hylobates syndactylus, as the name expresses, two of the digits regularly cohere; and this, as Mr. Blyth informs me, is occasionally the case with the digits of H. agilis, lar, and leuciscus.

203Brehm, 'Thierleben,' B. i. s. 80.

204"The Hand, its mechanism," &c. 'Bridgewater Treatise,' 1833, p. 38.

205Häckel has an excellent discussion on the steps by which man became a biped: 'Natürliche Schöpfungsgeschichte,' 1868, s. 507. Dr. Büchner ('Conférences sur la Théorie Darwinienne,' 1869, p. 135) has given good cases of the use of the foot as a prehensile organ by man; also on the manner of progression of the higher apes to which I allude in the following paragraph: see also Owen ('Anatomy of Vertebrates,' vol. iii. p. 71) on this latter subject.

206"On the Primitive Form of the Skull," translated in 'Anthropological Review,' Oct. 1868, p. 428. Owen ('Anatomy of Vertebrates,' vol. ii. 1866, p. 551) on the mastoid processes in the higher apes.

207'Die Grenzen der Thierwelt, eine Betrachtung zu Darwin's Lehre,' 1868, s. 51.

208Dujardin, 'Annales des Sc. Nat.' 3rd series, Zoolog. tom. xiv. 1850, p. 203. See also Mr. Lowne, 'Anatomy and Phys. of the Musca vomitoria,' 1870, p. 14. My son, Mr. F. Darwin, dissected for me the cerebral ganglia of the Formica rufa.

209'Philosophical Transactions,' 1869, p. 513.

210Quoted in C. Vogt's 'Lectures on Man,' Eng. translat. 1864, p. 88, 90. Prichard, 'Phys. Hist. of Mankind,' vol. i. 1838, p. 305.

211'Comptes Rendus des Séances,' &c. June 1, 1868.

212'The Variation of Animals and Plants under Domestication,' vol. ii. p. 124-129.

213Schaaffhausen gives from Blumenbach and Busch, the cases of the spasms and cicatrix, in 'Anthropolog. Review,' Oct. 1868, p. 420. Dr. Jarrold ('Anthropologia,' 1808, p. 115, 116) adduces from Camper and from his own observations, cases of the modification of the skull from the head being fixed in an unnatural position. He believes that certain trades, such as that of a shoemaker, by causing the head to be habitually held forward, makes the forehead more rounded and prominent.

214'Variation of Animals,' &c., vol. i. p. 117 on the elongation of the skull; p. 119, on the effect of the lopping of one ear.

215Quoted by Schaaffhausen, in 'Anthropolog. Review,' Oct. 1868, p. 419.

216Owen, 'Anatomy of Vertebrates,' vol. iii. p. 619.

217Isidore Geoffroy St.-Hilaire remarks ('Hist. Nat. Générale,' tom. ii. 1859, p. 215-217) on the head of man being covered with long hair; also on the upper surfaces of monkeys and of other mammals being more thickly clothed than the lower surfaces. This has likewise been observed by various authors. Prof. P. Gervais ('Hist. Nat. des Mammifères,' tom. i. 1854, p. 28), however, states that in the Gorilla the hair is thinner on the back, where it is partly rubbed off, than on the lower surface.

218Mr. St. George Mivart, 'Proc. Zoolog. Soc.' 1865, p. 562, 583. Dr. J. E. Gray, 'Cat. Brit. Mus.: Skeletons.' Owen, 'Anatomy of Vertebrates,' vol. ii. p. 517. Isidore Geoffroy, 'Hist. Nat. Gén.' tom. ii. p. 244.

219'The Variation of Animals and Plants under Domestication,' vol. ii. p. 280, 282.

220'Primeval Man,' 1869, p. 66.

221'Anthropological Review,' May, 1864, p. clviii.

222After a time the members or tribes which are absorbed into another tribe assume, as Mr. Maine remarks ('Ancient Law,' 1861, p. 131), that they are the co-descendants of the same ancestors.

223Morlot, 'Soc. Vaud. Sc. Nat.' 1860, p. 294.

224I have given instances in my 'Variation of Animals under Domestication,' vol. ii. p. 196.

225See a remarkable series of articles on Physics and Politics in the 'Fortnightly Review,' Nov. 1867; April 1, 1868; July 1, 1869.

226'Origin of Civilisation,' 1870, p. 265.

227Mr. Wallace gives cases in his 'Contributions to the Theory of Natural Selection,' 1870, p. 354.

228'Ancient Law,' 1861, p. 22. For Mr. Bagehot's remarks, 'Fortnightly Review,' April 1, 1868, p. 452.

229'The Variation of Animals and Plants under Domestication,' vol. i. p. 309.

230'Fraser's Magazine,' Sept. 1868, p. 353. This article seems to have struck many persons, and has given rise to two remarkable essays and a rejoinder in the 'Spectator,' Oct. 3rd and 17th 1868. It has also been discussed in the 'Q. Journal of Science,' 1869, p. 152, and by Mr. Lawson Tait in the 'Dublin Q. Journal of Medical Science,' Feb. 1869, and by Mr. E. Ray

Lankester in his 'Comparative Longevity,' 1870, p. 128. Similar views appeared previously in the 'Australasian,' July 13, 1867. I have borrowed ideas from several of these writers.

231For Mr. Wallace, see 'Anthropolog. Review,' as before cited. Mr. Galton in 'Macmillan's Magazine,' Aug. 1865, p. 318; also his great work, 'Hereditary Genius,' 1870.

232'Hereditary Genius,' 1870, p. 132-140.

233See the fifth and sixth columns, compiled from good authorities, in the table given in Mr. E. R. Lankester's 'Comparative Longevity,' 1870, p. 115.

234'Hereditary Genius,' 1870, p. 330.

235'Origin of Species' (fifth edition, 1869), p. 104.

236'Hereditary Genius,' 1870, p. 347.

237E. Ray Lankester, 'Comparative Longevity,' 1870, p. 115. The table of the intemperate is from Nelson's 'Vital Statistics.' In regard to profligacy, see Dr. Farr, "Influence of Marriage on Mortality," 'Nat. Assoc. for the Promotion of Social Science,' 1858.

238'Fraser's Magazine,' Sept. 1868, p. 353. 'Macmillan's Magazine,' Aug. 1865, p. 318. The Rev. F. W. Farrar ('Fraser's Mag.,' Aug. 1870, p. 264) takes a different view.

239"On the Laws of the Fertility of Women," in 'Transact. Royal Soc.' Edinburgh, vol. xxiv. p. 287. See, also, Mr. Galton, 'Hereditary Genius,' p. 352-357, for observations to the above effect.

240'Tenth Annual Report of Births, Deaths, &c., in Scotland,' 1867, p. xxix.

241These quotations are taken from our highest authority on such questions, namely, Dr. Farr, in his paper "On the Influence of Marriage on the Mortality of the French People," read before the Nat. Assoc. for the Promotion of Social Science, 1858.

242Dr. Farr, ibid. The quotations given below are extracted from the same striking paper.

243I have taken the mean of the quinquennial means, given in 'The Tenth Annual Report of Births, Deaths, &c., in Scotland,' 1867. The quotation from Dr. Stark is copied from an article in the 'Daily News,' Oct. 17th, 1868, which Dr. Farr considers very carefully written.

244See the ingenious and original argument on this subject by Mr. Galton, 'Hereditary Genius,' p. 340-342.

245Mr. Greg, 'Fraser's Magazine,' Sept. 1868, p. 357.

246'Hereditary Genius,' 1870, p. 357-359. The Rev. F. H. Farrar ('Fraser's Mag.', Aug. 1870, p. 257) advances arguments on the other side. Sir C. Lyell had already ('Principles of Geology,' vol. ii. 1868, p. 489) called attention, in a striking passage, to the evil influence of the Holy Inquisition in having lowered, through selection, the general standard of intelligence in Europe.

247Mr. Galton, 'Macmillan's Magazine,' August, 1865, p. 325. See, also, 'Nature,' "On Darwinism and National Life," Dec. 1869, p. 184.

248'Last Winter in the United States,' 1868, p. 29.

249'On the Origin of Civilisation,' 'Proc. Ethnological Soc.' Nov. 26, 1867.

250'Primeval Man,' 1869.

251'Royal Institution of Great Britain,' March 15, 1867. Also, 'Researches into the Early History of Mankind,' 1865.

252'Primitive Marriage,' 1865. See, likewise, an excellent article, evidently by the same author, in the 'North British Review,' July, 1869. Also, Mr. L. H. Morgan, "A Conjectural Solution of the Origin of the Class. System of Relationship," in 'Proc. American Acad. of Sciences,' vol. vii. Feb. 1868. Prof. Schaaffhausen ('Anthropolog. Review,' Oct. 1869, p. 373) remarks on "the vestiges of human sacrifices found both in Homer and the Old Testament."

253Sir J. Lubbock, 'Prehistoric Times,' 2nd edit. 1869, chap. xv. and xvi. et passim.

254Dr. F. Müller has made some good remarks to this effect in the 'Reise der Novara: Anthropolog. Theil,' Abtheil. iii. 1868, s. 127.

255Isidore Geoffroy St.-Hilaire gives a detailed account of the position assigned to man by various naturalists in their classifications: 'Hist. Nat. Gén.' tom. ii. 1859, p. 170-189.

256See the very interesting article, "L'Instinct chez les Insectes," by M. George Pouchet, 'Revue des Deux Mondes,' Feb. 1870, p. 682.

257Westwood, 'Modern Class. of Insects,' vol. ii. 1840, p. 87.

258'Proc. Zoolog. Soc.' 1869, p. 4.

259'Evidence as to Man's Place in Nature,' 1863, p. 70, et passim.

260Isid. Geoffroy, 'Hist. Nat. Gén.' tom. ii. 1859, p. 217.

261"Ueber die Richtung der Haare," &c., Müller's 'Archiv für Anat. und Phys.' 1837, s. 51.

262On the hair in Hylobates, see 'Nat. Hist. of Mammals,' by C. L. Martin, 1841, p. 415. Also, Isid. Geoffroy on the American monkeys and other kinds, 'Hist. Nat. Gén.' vol. ii. 1859, p. 216, 243. Eschricht, ibid. s. 46, 55, 61. Owen, 'Anat. of Vertebrates,' vol. iii. p. 619. Wallace, 'Contributions to the Theory of Natural Selection,' 1870. p. 344.

263'Origin of Species,' 5th edit. 1869, p. 194. 'The Variation of Animals and Plants under Domestication,' vol. ii. 1868, p. 348.

264'An Introduction to the Classification of Animals,' 1869, p. 99.

265This is nearly the same classification as that provisionally adopted by Mr. St. George Mivart ('Transact. Philosoph. Soc.' 1867, p. 300), who, after separating the Lemuridæ, divides the remainder of the Primates into the Hominidæ, the Simiadæ answering to the Catarhines, the Cebidæ, and the Hapalidæ,—these two latter groups answering to the Platyrhines.

266'Transact. Zoolog. Soc.' vol. vi. 1867, p. 214.

267Mr. St. G. Mivart, 'Transact. Phil. Soc.' 1867, p. 410.

268Messrs. Murie and Mivart on the Lemuroidea. 'Transact. Zoolog. Soc.' vol. vii. 1869, p. 5.

269Häckel has come to this same conclusion. See 'Ueber die Entstehung des Menschengeschlechts,' in Virchow's 'Sammlung. gemein. wissen. Vorträge,' 1868, s. 61. Also his 'Natürliche Schöpfungsgeschichte,' 1868, in which he gives in detail his views on the genealogy of man.

270'Anthropological Review,' April, 1867, p. 236.

271'Elements of Geology,' 1865, p. 583-585. 'Antiquity of Man', 1863; p. 145.

272'Man's Place in Nature,' p. 105.

273Elaborate tables are given in his 'Generelle Morphologie' (B. ii. s. cliii. and s. 425); and with more especial reference to man in his 'Natürliche Schöpfungsgeschichte,' 1868. Prof. Huxley, in reviewing this latter work ('The Academy,' 1869, p. 42) says, that he considers the phylum or lines of descent of the Vertebrata to be admirably discussed by Häckel, although he differs on some points. He expresses, also, his high estimate of the value of the general tenor and spirit of the whole work.

274'Palæontology,' 1860, p. 199.

275I had the satisfaction of seeing, at the Falkland Islands, in April, 1833, and therefore some years before any other naturalist, the locomotive larvæ of a compound Ascidian, closely allied to, but apparently generically distinct from, Synoicum. The tail was about five times as long as the oblong head, and terminated in a very fine filament. It was plainly divided, as sketched by me under a simple microscope, by transverse opaque partitions, which I presume represent the great cells figured by Kowalevsky. At an early stage of development the tail was closely coiled round the head of the larva.

276'Mémoires de l'Acad. des Sciences de St. Pétersbourg,' tom. x. No. 15, 1866.

277This is the conclusion of one of the highest authorities in comparative anatomy, namely, Prof. Gegenbaur: 'Grundzüge der vergleich. Anat.' 1870, s. 876. The result has been arrived at chiefly from the study of the Amphibia; but it appears from the researches of Waldeyer (as quoted in Humphry's 'Journal of Anat. and Phys.' 1869, p. 161), that the sexual organs of even "the higher vertebrata are, in their early condition, hermaphrodite." Similar views have long been held by some authors, though until recently not well based.

278The male Thylacinus offers the best instance. Owen, 'Anatomy of Vertebrates,' vol. iii. p. 771.

279Serranus is well known often to be in an hermaphrodite condition; but Dr. Günther informs me that he is convinced that this is not its normal state. Descent from an ancient androgynous prototype would, however, naturally favour and explain, to a certain extent, the recurrence of this condition in these fishes.

280Mr. Lockwood believes (as quoted in 'Quart. Journal of Science,' April, 1868, p. 269), from what he has observed of the development of Hippocampus, that the walls of the abdominal pouch of the male in some way afford nourishment. On male fishes hatching the ova in their mouths, see a very interesting paper by Prof. Wyman, in 'Proc. Boston Soc. of Nat. Hist.' Sept. 15, 1857; also Prof. Turner, in 'Journal of Anat. and Phys.' Nov. 1, 1866, p. 78. Dr. Günther has likewise described similar cases.

281All vital functions tend to run their course in fixed and recurrent periods, and with tidal animals the periods would probably be lunar; for such animals must have been left dry or covered deep with water,—supplied with copious food or stinted,—during endless generations, at regular lunar intervals. If then the Vertebrata are descended from an animal allied to the existing tidal Ascidians, the mysterious fact, that with the higher and now terrestrial Vertebrata, not to mention other classes, many normal and abnormal vital processes run their course according to lunar periods, is rendered intelligible. A recurrent period, if approximately of the right duration, when once gained, would not, as far as we can judge, be liable to be changed; consequently it might be thus transmitted during almost any number of generations. This conclusion, if it could be proved sound, would be curious; for we should then see that the period of gestation in each mammal, and the hatching of each bird's eggs, and many other vital processes, still betrayed the primordial birthplace of these animals.

282'History of India,' 1841, vol. i. p. 323. Father Ripa makes exactly the same remark with respect to the Chinese.

283A vast number of measurements of Whites, Blacks, and Indians, are given in the 'Investigations in the Military and Anthropolog. Statistics of American Soldiers,' by B. A. Gould, 1869, p. 298-358; on the capacity of the lungs, p. 471. See also the numerous and valuable tables, by Dr. Weisbach, from the observations of Dr. Scherzer and Dr. Schwarz, in the 'Reise der Novara: Anthropolog. Theil,' 1867.

284See, for instance, Mr. Marshall's account of the brain of a Bush-woman, in 'Phil. Transact.' 1864, p. 519.

285Wallace, 'The Malay Archipelago,' vol. ii. 1869, p. 178.

286With respect to the figures in the famous Egyptian caves of Abou-Simbel, M. Pouchet says ('The Plurality of the Human Races,' Eng. translat. 1864, p. 50), that he was far from finding recognisable representations of the dozen or more nations which some authors believe that they can recognise. Even some of the most strongly-marked races cannot be identified with that degree of unanimity which might have been expected from what has been written on the subject. Thus Messrs. Nott and Gliddon ('Types of Mankind,' p. 148) state that Rameses II., or the Great, has features superbly European; whereas Knox, another firm believer in the specific distinction of the races of man ('Races of Man,' 1850, p. 201), speaking of young Memnon (the same person with Rameses II., as I am informed by Mr. Birch) insists in the strongest manner that he is identical in character with the Jews of Antwerp. Again, whilst looking in the British Museum with two competent judges, officers of the establishment, at the statue of Amunoph III., we agreed that he had a strongly negro cast of features; but Messrs. Nott and Gliddon (ibid. p. 146, fig. 53) describe him as "a hybrid, but not of negro intermixture."

287As quoted by Nott and Gliddon, 'Types of Mankind,' 1854, p. 439. They give also corroborative evidence; but C. Vogt thinks that the subject requires further investigation.

288"Diversity of Origin of the Human Races," in the 'Christian Examiner,' July, 1850.

289'Transact. B. Soc. of Edinburgh,' vol. xxii. 1861, p. 567.

290'On the Phenomena of Hybridity in the Genus Homo,' Eng. translat. 1864.

291See the interesting letter by Mr. T. A. Murray, in the 'Anthropolog. Review,' April, 1868, p. liii. In this letter Count Strzelecki's statement, that Australian women who have borne children to a white man are afterwards sterile with their own race, is disproved. M. A. de Quatrefages has also collected ('Revue des Cours Scientifiques,' March, 1869, p. 239) much evidence that Australians and Europeans are not sterile when crossed.

292'An Examination of Prof. Agassiz's Sketch of the Nat. Provinces of the Animal World,' Charleston, 1855, p. 44.

293'Military and Anthropolog. Statistics of American Soldiers,' by B. A. Gould, 1869, p. 319.

294'The Variation of Animals and Plants under Domestication,' vol. ii. p. 109. I may here remind the reader that the sterility of species when crossed is not a specially-acquired quality; but, like the incapacity of certain trees to be grafted together, is incidental on other acquired differences. The nature of these differences is unknown, but they relate more especially to the reproductive system, and much less to external structure or to ordinary differences in constitution. One important element in the sterility of crossed species apparently lies in one or both having been long habituated to fixed conditions; for we know that changed conditions have a special influence on the reproductive system, and we have good reason to believe (as before remarked) that the fluctuating conditions of domestication tend to eliminate that sterility which is so general with species in a natural state when crossed. It has elsewhere been shewn by me (ibid. vol. ii. p. 185, and 'Origin of Species,' 5th edit. p. 317) that the sterility of crossed species has not been acquired through natural selection: we can see that when two forms have already been rendered very sterile, it is scarcely possible that their sterility should be augmented by the preservation or survival of the more and more sterile individuals; for as the sterility increases fewer and fewer offspring will be produced from which to breed, and at last only single individuals will be produced, at the rarest intervals. But there is even a higher grade of sterility than this. Both Gärtner and Kölreuter have proved that in genera of plants including numerous species, a series can be formed from species which when crossed yield fewer and fewer seeds, to species which never produce a single seed, but yet are affected by the pollen of the other species, for the germen swells. It is here manifestly impossible to select the more sterile individuals, which have already ceased to yield seeds; so that the acme of sterility, when the germen alone is affected, cannot be gained through selection. This acme, and no doubt the other grades of sterility, are the incidental results of certain unknown differences in the constitution of the reproductive system of the species which are crossed.

295'The Variation of Animals,' &c., vol. ii. p. 92.

296M. de Quatrefages has given ('Anthropolog. Review,' Jan. 1869, p. 22) an interesting account of the success and energy of the Paulistas in Brazil, who are a much crossed race of Portuguese and Indians, with a mixture of the blood of other races.

297For instance with the aborigines of America and Australia. Prof. Huxley says ('Transact. Internat. Congress of Prehist. Arch.' 1868. p. 105) that the skulls of many South Germans and Swiss are "as short and as broad as those of the Tartars," &c.

298See a good discussion on this subject in Waitz, 'Introduct. to Anthropology,' Eng. translat. 1863, p. 198-208, 227. I have taken some of the above statements from H. Tuttle's 'Origin and Antiquity of Physical Man,' Boston, 1866, p. 35.

299Prof. Nägeli has carefully described several striking cases in his 'Botanische Mittheilungen,' B. ii. 1866, s. 294-369. Prof. Asa Gray has made analogous remarks on some intermediate forms in the Compositæ of N. America.

300'Origin of Species,' 5th edit. p. 68.

301See Prof. Huxley to this effect in the 'Fortnightly Review,' 1865, p. 275.

302'Lectures on Man,' Eng. translat. 1864, p. 468.

303'Die Racen des Schweines,' 1860, s. 46. 'Vorstudien für Geschichte, &c., Schweineschädel,' 1864, s. 104. With respect to cattle, see M. de Quatrefages, 'Unité de l'Espèce Humaine,' 1861, p. 119.

304Tylor's 'Early History of Mankind,' 1865; for the evidence with respect to gesture-language, see p. 54. Lubbock's 'Prehistoric Times,' 2nd edit. 1869.

305'The Primitive Inhabitants of Scandinavia,' Eng. translat. edited by Sir J. Lubbock, 1868, p. 104.

306Hodder M. Westropp, on Cromlechs, &c., 'Journal of Ethnological Soc.' as given in 'Scientific Opinion,' June 2nd. 1869, p. 3.

307'Journal of Researches: Voyage of the "Beagle,"' p. 46.

308'Prehistoric Times,' 1869, p. 574.

309Translation in 'Anthropological Review,' Oct. 1868, p. 431.

310'Transact. Internat. Congress of Prehistoric Arch.' 1868, p. 172-175. See also Broca (translation) in 'Anthropological Review,' Oct. 1868, p. 410.

311Dr. Gerland, 'Ueber das Aussterben der Naturvölker,' 1868, s. 82.

312Gerland (ibid. s. 12) gives facts in support of this statement.

313See remarks to this effect in Sir H. Holland's 'Medical Notes and Reflections,' 1839, p. 390.

314I have collected ('Journal of Researches, Voyage of the "Beagle,"' p. 435) a good many cases bearing on this subject: see also Gerland, ibid. s. 8. Poeppig speaks of the "breath of civilisation as poisonous to savages."

315Sproat, 'Scenes and Studies of Savage Life,' 1868, p. 284.

316Bagehot, "Physics and Politics," 'Fortnightly Review,' April 1, 1868, p. 455.

317"On Anthropology," translation, 'Anthropolog. Review,' Jan. 1868, p. 38.

318'The Annals of Rural Bengal,' 1868, p. 134.

319'The Variation of Animals and Plants under Domestication,' vol. ii. p. 95.

320Pallas, 'Act. Acad. St. Petersburgh,' 1780, part ii. p. 69. He was followed by Rudolphi, in his 'Beyträge zur Anthropologie,' 1812. An excellent summary of the evidence is given by Godron, 'De l'Espèce,' 1859, vol. ii. p. 246, &c.

321Sir Andrew Smith, as quoted by Knox, 'Races of Man,' 1850, p. 473.

322See De Quatrefages on this head, 'Revue des Cours Scientifiques,' Oct. 17, 1868, p. 731.

323Livingstone's 'Travels and Researches in S. Africa,' 1857, p. 338, 329. D'Orbigny, as quoted by Godron, 'De l'Espèce,' vol. ii. p. 266.

324See a paper read before the Royal Soc. in 1813, and published in his Essays in 1818. I have given an account of Dr. Wells' views in the Historical Sketch (p. xvi) to my 'Origin of Species.' Various cases of colour correlated with constitutional peculiarities are given in my 'Variation of Animals under Domestication,' vol. ii. p. 227, 335.

325See, for instance, Nott and Gliddon, 'Types of Mankind,' p. 68.

326Major Tulloch, in a paper read before the Statistical Society, April 20th, 1840, and given in the 'Athenæum,' 1840, p. 353.

327'The Plurality of the Human Race' (translat.), 1864, p. 60.

328Quatrefages, 'Unité de l'Espèce Humaine,' 1861, p. 205. Waitz, 'Introduct. to Anthropology,' translat. vol. i. 1863, p. 124. Livingstone gives analogous cases in his 'Travels.'

329In the spring of 1862 I obtained permission from the Director-General of the Medical department of the Army, to transmit to the surgeons of the various regiments on foreign service a blank table, with the following appended remarks, but I have received no returns. "As several well-marked cases have been recorded with our domestic animals of a relation between the colour of the dermal appendages and the constitution; and it being notorious that there is some limited degree of relation between the colour of the races of man and the climate inhabited by them; the following investigation seems worth consideration. Namely, whether there is any relation in Europeans between the colour of their hair, and their liability to the diseases of tropical countries. If the surgeons of the several regiments, when stationed in unhealthy tropical districts, would be so good as first to count, as a standard of comparison, how many men, in the force whence the sick are drawn, have dark and light-coloured hair, and hair of intermediate or doubtful tints; and if a similar account were kept by the same medical gentlemen, of all the men who suffered from malarious and yellow fevers, or from dysentery, it would soon be apparent, after some thousand cases had been tabulated, whether there exists any relation between the colour of the hair and constitutional liability to tropical diseases. Perhaps no such relation would be discovered, but the investigation is well worth making. In case any positive result were obtained, it might be of some practical use in selecting men for any particular service. Theoretically the result would be of high interest, as indicating one means by which a race of men inhabiting from a remote period an unhealthy tropical climate, might have become dark-coloured by the better preservation of dark-haired or dark-complexioned individuals during a long succession of generations."

330'Anthropological Review,' Jan. 1866, p. xxi.

331See, for instance, Quatrefages ('Revue des Cours Scientifiques,' Oct. 10, 1868, p. 724) on the effects of residence in Abyssinia and Arabia, and other analogous cases. Dr. Rolle ('Der Mensch, seine Abstammung,' &c., 1865, s. 99) states, on the authority of Khanikof, that the greater number of German families settled in Georgia, have acquired in the course of two generations dark hair and eyes. Mr. D. Forbes informs me that the Quichuas in the Andes vary greatly in colour, according to the position of the valleys inhabited by them.

332Harlan, 'Medical Researches,' p. 532. Quatrefages ('Unité de l'Espèce Humaine,' 1861, p. 128) has collected much evidence on this head.

333See Prof. Schaaffhausen, translat. in 'Anthropological Review,' Oct. 1868, p. 429.

334Mr. Catlin states ('N. American Indians,' 3rd edit. 1842, vol. i. p. 49) that in the whole tribe of the Mandans, about one in ten or twelve of the members of all ages and both sexes have bright silvery grey hair, which is hereditary. Now this hair is as coarse and harsh as that of a horse's mane, whilst the hair of other colours is fine and soft.

335On the odour of the skin, Godron, 'Sur l'Espèce,' tom. ii. p. 217. On the pores in the skin, Dr. Wilckens, 'Die Aufgaben der landwirth. Zootechnik,' 1869, s. 7.

336Westwood, 'Modern Class. of Insects,' vol. ii. 1810, p. 541. In regard to the statement about Tanais, mentioned below, I am indebted to Fritz Müller.

337Kirby and Spence, 'Introduction to Entomology,' vol. iii. 1826, p. 309.

338Even with those of plants in which the sexes are separate, the male flowers are generally mature before the female. Many hermaphrodite plants are, as first shewn by C. K. Sprengel, dichogamous; that is, their male and female organs are not ready at the same time, so that they cannot be self-fertilised. Now with such plants the pollen is generally mature in the same flower before the stigma, though there are some exceptional species in which the female organs are mature before the male.

339I have received information, hereafter to be given, to this effect with respect to poultry. Even with birds, such as pigeons, which pair for life, the female, as I hear from Mr. Jenner Weir, will desert her mate if he is injured or grows weak.

340On the Gorilla, Savage and Wyman, 'Boston Journal of Nat. Hist.' vol. v. 1845-47, p. 423. On Cynocephalus, Brehm, 'Illust. Thierleben,' B. i. 1864, s. 77. On Mycetes, Rengger, 'Naturgesch.: Säugethiere von Paraguay,' 1830, s. 14, 20. On Cebus, Brehm, ibid. s. 108.

341Pallas, 'Spicilegia Zoolog.' Fasc. xii. 1777, p. 29. Sir Andrew Smith, 'Illustrations of the Zoology of S. Africa,' 1849, pl. 29, on the Kobus. Owen, in his 'Anatomy of Vertebrates' (vol. iii. 1868, p. 633) gives a table incidentally showing which species of Antelopes pair and which are gregarious.

342Dr. Campbell, in 'Proc. Zoolog. Soc.' 1869, p. 138. See also an interesting paper, by Lieut. Johnstone, in 'Proc. Asiatic Soc. of Bengal,' May, 1868.

343'The Ibis,' vol. iii. 1861, p. 133, on the Progne Widow-bird. See also on the Vidua axillaris, ibid. vol. ii. 1860, p. 211. On the polygamy of the Capercailzie and Great Bustard, see L. Lloyd, 'Game Birds of Sweden,' 1867, p. 19, and 182. Montagu and Selby speak of the Black Grouse as polygamous and of the Red Grouse as monogamous.

344The Rev. E. S. Dixon, however, speaks positively ('Ornamental Poultry,' 1848, p. 76) about the eggs of the guinea-fowl being infertile when more than one female is kept with the same male.

345Noel Humphreys, 'River Gardens,' 1857.

346Kirby and Spence, 'Introduction to Entomology,' vol. iii. 1826, p. 342.

347One parasitic Hymenopterous insect (Westwood, 'Modern Class. of Insects,' vol. ii, p. 160) forms an exception to the rule, as the male has rudimentary wings, and never quits the cell in which it is born, whilst the female has well-developed wings. Audouin believes that the females are impregnated by the males which are born in the same cells with them; but it is much more probable that the females visit other cells, and thus avoid close inter-breeding. We shall hereafter meet with a few exceptional cases, in various classes, in which the female, instead of the male, is the seeker and wooer.

348'Essays and Observations,' edited by Owen, vol. i. 1861, p. 194.

349Prof. Sachs ('Lehrbuch der Botanik,' 1870, s. 633) in speaking of the male and female reproductive cells, remarks, "verhält sich die eine bei der Vereinigung activ, ... die andere erscheint bei der Vereinigung passiv."

350'Reise der Novara: Anthropolog. Theil,' 1867, s. 216-269. The results were calculated by Dr. Weisbach from measurements made by Drs. K. Scherzer and Schwarz. On the greater variability of the males of domesticated animals, see my 'Variation of Animals and Plants under Domestication,' vol. ii. 1868, p. 75.

351'Proceedings Royal Soc.' vol. xvi. July, 1868, p. 519 and 524.

352'Proc. Royal Irish Academy,' vol. x. 1868, p. 123.

353'Massachusetts Medical Soc.' vol. ii. No. 3, 1808, p. 9.

354'The Variation of Animals and Plants under Domestication,' vol. ii. 1868, p. 75. In the last chapter but one, the provisional hypothesis of pangenesis, above alluded to, is fully explained.

355These facts are given on the high authority of a great breeder, Mr. Teebay, in Tegetmeier's 'Poultry Book,' 1868, p. 158. On the characters of chickens of different breeds, and on the breeds of the pigeon, alluded to in the above paragraph, see 'Variation of Animals,' &c., vol. i. p. 160, 249; vol. ii. p. 77.

356'Novæ species Quadrupedum e Glirium ordine,' 1778, p. 7. On the transmission of colour by the horse, see 'Variation of Animals, &c. under Domestication,' vol. i. p. 21. Also vol. ii. p. 71, for a general discussion on Inheritance as limited by Sex.

357Dr. Chapuis, 'Le Pigeon Voyageur Belge,' 1865, p. 87. Boitard et Corbié, 'Les Pigeons de Volière,' &c., 1824, p. 173.

358References are given in my 'Variation of Animals under Domestication,' vol. ii. p. 72.

359I am much obliged to Mr. Cupples for having made enquiries for me in regard to the Roebuck and Red Deer of Scotland from Mr. Robertson, the experienced head-forester to the Marquis of Breadalbane. In regard to Fallow-deer, I am obliged to Mr. Eyton and others for information. For theCervus alces of N. America, see 'Land and Water,' 1868, p. 221 and 254; and for the C. Virginianus and strongyloceros of the same continent, see J. D. Caton, in 'Ottawa Acad. of Nat. Sc.' 1868, p. 13. For Cervus Eldi of Pegu, see Lieut. Beavan, 'Proc. Zoolog. Soc.' 1867, p. 762.

360Antilocapra Americana. Owen, 'Anatomy of Vertebrates,' vol. iii. p. 627.

361I have been assured that the horns of the sheep in North Wales can always be felt, and are sometimes even an inch in length, at birth. With cattle Youatt says ('Cattle,' 1834, p. 277) that the prominence of the frontal bone penetrates the cutis at birth, and that the horny matter is soon formed over it.

362I am greatly indebted to Prof. Victor Carus for having made inquiries for me, from the highest authorities, with respect to the merino sheep of Saxony. On the Guinea coast of Africa there is a breed of sheep in which, as with merinos, the rams alone bear horns; and Mr. Winwood Reade informs me that in the one case observed, a young ram born on Feb. 10th first showed horns on March 6th, so that in this instance the development of the horns occurred at a later period of life, conformably with our rule, than in the Welsh sheep, in which both sexes are horned.

363In the common peacock (Pavo cristatus) the male alone possesses spurs, whilst both sexes of the Java peacock (P. muticus) offer the unusual case of being furnished with spurs. Hence I fully expected that in the latter species they would have been developed earlier in life than in the common peacock; but M. Hegt of Amsterdam informs me, that with young birds of the previous year, belonging to both species, compared on April 23rd, 1869, there was no difference in the development of the spurs. The spurs, however, were as yet represented merely by slight knobs or elevations. I presume that I should have been informed if any difference in the rate of development had subsequently been observed.

364In some other species of the Duck Family the speculum in the two sexes differs in a greater degree; but I have not been able to discover whether its full development occurs later in life in the males of such species, than in the male of the common duck, as ought to be the case according to our rule. With the allied Mergus cucullatus we have, however, a case of this kind: the two sexes differ conspicuously in general plumage, and to a considerable degree in the speculum, which is pure white in the male and greyish-white in the female. Now the young males at first resemble, in all respects, the female, and have a greyish-white speculum, but this becomes pure white at an earlier age than that at which the adult male acquires his other more strongly-marked sexual differences in plumage: see Audubon, 'Ornithological Biography,' vol. iii. 1835, p. 249-250.

365'Das Ganze der Taubenzucht,' 1837, s. 21, 24. For the case of the streaked pigeons, see Dr. Chapuis, 'Le Pigeon Voyageur Belge.' 1865, p. 87.

366For full particulars and references on all these points respecting the several breeds of the Fowl, see 'Variation of Animals and Plants under Domestication,' vol. i. p. 250, 256. In regard to the higher animals, the sexual differences which have arisen under domestication are described in the same work under the head of each species.

367'Twenty-ninth Annual Report of the Registrar-General for 1866.' In this report (p. xii) a special decennial table is given.

368For Norway and Russia, see abstract of Prof. Faye's researches, in 'British and Foreign Medico-Chirurg. Review,' April, 1867, p. 343, 345. For France, the 'Annuaire pour l'An 1867.' p. 213.

369In regard to the Jews, see M. Thury, 'La Loi de Production des Sexes,' 1863, p. 25.

370Babbage, 'Edinburgh Journal of Science,' 1829, vol. i. p. 88; also p. 90, on still-born children. On illegitimate children in England, see 'Report of Registrar-General for 1866,' p. xv.

371'British and Foreign Medico-Chirurg. Review,' April, 1867, p. 343. Dr. Stark also remarks ('Tenth Annual Report of Births, Deaths, &c., in Scotland,' 1867, p. xxviii) that "These examples may suffice to shew that, at almost every stage of life, the males in Scotland have a greater liability to death and a higher death-rate than the females. The fact, however, of this peculiarity being most strongly developed at that infantile period of life when the dress, food, and general treatment of both sexes are alike, seems to prove that the higher male death-rate is an impressed, natural, and constitutional peculiarity due to sex alone."

372With the savage Guaranys of Paraguay, according to the accurate Azara ('Voyages dans l'Amérique mérid.' tom. ii. 1809, p. 60, 179), the women in proportion to the men are as 14 to 13.

373Leuckart in Wagner, 'Handwörterbuch der Phys.' B. iv. 1853, s. 774.

374Anthropological Review, April, 1870, p. cviii.

375During the last eleven years a record has been kept of the number of mares which have proved barren or prematurely slipped their foals; and it deserves notice, as shewing how infertile these highly-nurtured and rather closely-interbred animals have become, that not far from one-third of the mares failed to produce living foals. Thus during 1866, 809 male colts and 816 female colts were born, and 743 mares failed to produce offspring. During 1867, 836 males and 902 females were born, and 794 mares failed.

376I am much indebted to Mr. Cupples for having procured for me the above returns from Scotland, as well as some of the following returns on cattle. Mr. R. Elliot, of Laighwood, first called my attention to the premature deaths of the males,—a statement subsequently confirmed by Mr. Aitchison and others. To this latter gentleman, and to Mr. Payan, I owe my thanks for the larger returns on sheep.

377Bell, 'History of British Quadrupeds,' p. 100.

378'Illustrations of the Zoology of S. Africa,' 1849, pl. 29.

379Brehm ('Illust. Thierleben,' B. iv. s. 990) comes to the same conclusion.

380On the authority of L. Lloyd, 'Game Birds of Sweden,' 1867, p. 12, 132.

381'Nat. Hist. of Selbourne,' letter xxix. edit. of 1825, vol. i. p. 139.

382Mr. Jenner Weir received similar information, on making enquiries during the following year. To shew the number of chaffinches caught, I may mention that in 1869 there was a match between two experts; and one man caught in a day 62, and another 40, male chaffinches. The greatest number ever caught by one man in a single day was 70.

383'Ibis,' vol. ii. p. 260, as quoted in Gould's 'Trochilidæ,' 1861, p. 52. For the foregoing proportions, I am indebted to Mr. Salvin for a table of his results.

384'Ibis,' 1860, p. 137; and 1867, p. 369.

385'Ibis,' 1862, p. 137.

386Leuckart quotes Bloch (Wagner, 'Handwörterbuch der Phys.' B. iv. 1853, s. 775), that with fish there are twice as many males as females.

387Quoted in the 'Farmer,' March 18, 1869, p. 369.

388'The Stormontfield Piscicultural Experiments,' 1866, p. 23. The 'Field' newspaper, June 29th, 1867.

389'Land and Water,' 1868, p. 41.

390Yarrell, 'Hist. British Fishes,' vol. i. 1836, p. 307; on the Cyprinus carpio, p. 331; on the Tinca vulgaris, p. 331; on the Abramis brama, p. 336. See, for the minnow (Leuciscus phoxinus), 'Loudon's Mag. of Nat. Hist.' vol. v. 1832, p. 682.

391Leuckart quotes Meinecke (Wagner, 'Handwörterbuch der Phys.' B. iv. 1853, s. 775) that with Butterflies the males are three or four times as numerous as the females.

392'The Naturalist on the Amazons,' vol. ii. 1863, p. 228, 347.

393Four of these cases are given by Mr. Trimen in his 'Rhopalocera Africæ Australis.'

394Quoted by Trimen, 'Transact. Ent. Soc.' vol. v. part iv. 1866, p. 330.

395'Transact. Linn. Soc.' vol. xxv. p. 37.

396'Proc. Entomolog. Soc.' Feb. 17th, 1868.

397Quoted by Dr. Wallace in 'Proc. Ent. Soc.' 3rd series, vol. v. 1867, p. 487.

398Blanchard, 'Metamorphoses, Mœurs des Insectes,' 1868, p. 225-226.

399'Lepidopteren-Doubblettren Liste,' Berlin, No. x. 1866.

400This naturalist has been so kind as to send me some results from former years, in which the females seemed to preponderate; but so many of the figures were estimates, that I found it impossible to tabulate them.

401Günther's 'Record of Zoological Literature,' 1867, p. 260. On the excess of female Lucanus, ibid. p. 250. On the males of Lucanus in England, Westwood, 'Modern Class. of Insects,' vol. i. p. 187. On the Siagonium, ibid. p. 172.

402Walsh, in 'The American Entomologist,' vol. i. 1869, p. 103. F. Smith, 'Record of Zoological Literature,' 1867, p. 328.

403'Farm Insects,' p. 45-46.

404'Observations on N. American Neuroptera,' by H. Hagen and B. D. Walsh, 'Proc. Ent. Soc. Philadelphia,' Oct. 1863, p. 168, 223, 239.

405'Proc. Ent. Soc. London,' Feb. 17, 1868.

406Another great authority in this class, Prof. Thorell of Upsala ('On European Spiders,' 1869-70, part i. p. 205) speaks as if female spiders were generally commoner than the males.

407See, on this subject, Mr. Pickard-Cambridge, as quoted in 'Quarterly Journal of Science,' 1868, p. 429.

408I have often been struck with the fact, that in several species of Primula the seeds in the capsules which contained only a few were very much larger than the numerous seeds in the more productive capsules.

409'Principles of Biology,' vol. ii. 1867, chaps. ii.-xi.

410De l'Espèce et de la Class.' &c., 1869, p. 106.

411See, for instance, the account which I have given in my 'Journal of Researches,' 1845, p. 7.

412I have given ('Geolog. Observations on Volcanic Islands,' 1844, p. 53) a curious instance of the influence of light on the colours of a frondescent incrustation, deposited by the surf on the coast-rocks of Ascension, and formed by the solution of triturated sea-shells.

413'Facts and Arguments for Darwin,' English translat. 1869, p. 20. See the previous discussion on the olfactory threads. Sars has described a somewhat analogous case (as quoted in 'Nature,' 1870, p. 455) in a Norwegian crustacean, the Pontoporeia affinis.

414See Sir J. Lubbock in 'Annals. and Mag. of Nat. Hist.' vol. xi. 1853, pl. i. and x.; and vol. xii. (1853) pl. vii. See also Lubbock in 'Transact. Ent. Soc.' vol. iv. new series, 1856-1858, p. 8. With respect to the zigzagged antennæ mentioned below, see Fritz Müller, 'Facts and Arguments for Darwin' 1869, p. 40, foot-note.

415See a paper by Mr. C. Spence Bate, with figures, in 'Proc. Zoolog. Soc.' 1868, p. 363; and on the nomenclature of the genus, ibid. p. 585. I am greatly indebted to Mr. Spence Bate for nearly all the above statements with respect to the chelæ of the higher crustaceans.

416'Hist. Nat. des Crust.' tom. ii. 1837, p. 50.

417Fritz Müller, 'Facts and Arguments for Darwin,' 1869, p. 25-28.

418'Travels in the Interior of Brazil,' 1846, p. 111. I have given, in my 'Journal of Researches,' p. 463, an account of the habits of the Birgos.

419Mr. Ch. Fraser, in 'Proc. Zoolog. Soc.' 1869, p. 3. I am indebted to Mr. Bate for the statement from Dr. Power.

420Claus, 'Die freilebenden Copepoden,' 1863, s. 35.

421'Facts and Arguments,' &c., p. 79.

422'A History of the Spiders of Great Britain,' 1861-64. For the following facts, see p. 102, 77, 88.

423Aug. Vinson ('Aranéides des Iles de la Réunion,' pl. vi. figs. 1 and 2) gives a good instance of the small size of the male in Epeira nigra. In this species, as I may add, the male is testaceous and the female black with legs banded with red. Other even more striking cases of inequality in size between the sexes have been recorded ('Quarterly Journal of Science,' 1868, July, p. 429); but I have not seen the original accounts.

424Kirby and Spence, 'Introduction to Entomology,' vol. i. 1818, p. 280.

425Theridion (Asagena, Sund.) serratipes, 4-punctatum et guttatum; see Westring, in Kroyer, 'Naturhist. Tidskrift,' vol. iv. 1842-1843, p. 349; and vol. ii. 1846-1849, p. 342. See, also, for other species, 'Araneæ Svecicæ,' p. 184.

426Walckenaer et P. Gervais, 'Hist. Nat. des Insectes: Aptères,' tom. iv. 1847, p. 17, 19, 68.

427Sir J. Lubbock, 'Transact. Linnean Soc.' vol. xxv. 1866, p. 484. With respect to the Mutillidæ see Westwood, 'Modern Class. of Insects,' vol. ii. p. 213.

428These organs in the male often differ in closely-allied species, and afford excellent specific characters. But their importance, under a functional point of view, as Mr. E. MacLachlan has remarked to me, has probably been overrated. It has been suggested, that slight differences in these organs would suffice to prevent the intercrossing of well-marked varieties or incipient species, and would thus aid in their development. That this can hardly be the case, we may infer from the many recorded cases (see for instance, Bronn, 'Geschichte der Natur,' B. ii. 1843, s. 164; and Westwood, 'Transact. Ent. Soc.' vol. iii. 1842, p. 195) of distinct species having been observed in union. Mr. MacLachlan informs me (vide 'Stett. Ent. Zeitung,' 1867, s. 155) that when several species of Phryganidæ, which present strongly-pronounced differences of this kind, were confined together by Dr. Aug. Meyer, they coupled, and one pair produced fertile ova.

429'The Practical Entomologist,' Philadelphia, vol. ii. May, 1867, p. 88.

430Mr. Walsh, ibid. p. 107.

431'Modern Classification of Insects,' vol. ii. 1840, p. 206, 205. Mr. Walsh, who called my attention to this double use of the jaws, says that he has repeatedly observed this fact.

432We have here a curious and inexplicable case of dimorphism, for some of the females of four European species of Dytiscus, and of certain species of Hydroporus, have their elytra smooth; and no intermediate gradations between sulcated or punctured and quite smooth elytra have been observed. See Dr. H. Schaum, as quoted in the 'Zoologist,' vol. v.-vi. 1847-48, p. 1896. Also Kirby and Spence, 'Introduction to Entomology,' vol. iii. 1826, p. 305.

433Westwood, 'Modern Class.' vol. ii. p. 193. The following statement about Penthe, and others in inverted commas, are taken from Mr. Walsh, 'Practical Entomologist,' Philadelphia, vol. ii. p. 88.

434Kirby and Spence, 'Introduct.' &c., vol. iii. p. 332-336.

435'Insecta Maderensia,' 1854, p. 20.

436E. Doubleday, 'Annals and Mag. of Nat. Hist.' vol. i. 1848, p. 379. I may add that the wings in certain Hymenoptera (see Shuckard, 'Fossorial Hymenop.' 1837, p. 39-43) differ in neuration according to sex.

437H. W. Bates, in 'Journal of Proc. Linn. Soc.' vol. vi. 1862, p. 74. Mr. Wonfor's observations are quoted in 'Popular Science Review,' 1868, p. 343.

438Kirby and Spence, 'Introduction to Entomology,' vol. iii. p. 299.

439Robinet, 'Vers à Soie,' 1848, p. 207.

440'Transact. Ent. Soc.' 3rd series, vol. v. p. 486.

441'Journal of Proc. Ent. Soc.' Feb. 4th, 1867, p. lxxi.

442For this and other statements on the size of the sexes, see Kirby and Spence, ibid. vol. iii. p. 300; on the duration of life in insects, see p. 344.

443'Transact. Linnean Soc.' vol. xxvi. 1868, p. 296.

444'The Malay Archipelago,' vol. ii. 1869, p. 313.

445'Modern Classification of Insects,' vol. ii. 1840, p. 526.

446See Mr. B. T. Lowne's very interesting work, 'On the Anatomy of the Blow-Fly, Musca vomitoria,' 1870, p. 14.

447Westwood, 'Modern Class. of Insects,' vol. ii. p. 473.

448These particulars are taken from Westwood's 'Modern Class. of Insects,' vol. ii. 1840, p. 422. See, also, on the Fulgoridæ, Kirby and Spence, 'Introduct.' vol. ii. p. 401.

449'Zeitschrift für wissenschaft. Zoolog.' B. xvii. 1867, s. 152-158.

450I am indebted to Mr. Walsh for having sent me this extract from a 'Journal of the Doings of Cicada septemdecim,' by Dr. Hartman.

451L. Guilding, 'Transact. Linn. Soc.' vol. xv. p. 154.

452Köppen, as quoted in the 'Zoological Record,' for 1867, p. 460.

453Gilbert White, 'Nat. Hist. of Selborne,' vol. ii. 1825, p. 262.

454Harris, 'Insects of New England,' 1842, p. 128.

455'The Naturalist on the Amazons,' vol. i. 1863, p. 252. Mr. Bates gives a very interesting discussion on the gradations in the musical apparatus of the three families. See also Westwood, 'Modern Class.' vol. ii. p. 445 and 453.

456'Proc. Boston Soc. of Nat. Hist.' vol. xi. April, 1868.

457'Nouveau Manuel d'Anat. Comp.' (French translat.), tom. i. 1850 p. 567.

458'Zeitschrift für wissenschaft. Zoolog.' B. xvii. 1867, s. 117.

459Westwood, 'Modern Class. of Insects,' vol. i. p. 440.

460Westwood, 'Modern Class. of Insects,' vol. i. p. 453.

461Landois, ibid. s. 121, 122.

462Mr. Walsh also informs me that he has noticed that the female of the Platyphyllum concavum, "when captured makes a feeble grating noise by shuffling her wing-covers together."

463Landois, ibid. s. 113.

464'Insects of New England,' 1842, p. 133.

465Westwood, 'Modern Classification,' vol. i. p. 462.

466Westwood, ibid. vol. i. p. 453.

467Landois, ibid. s. 115, 116, 120, 122.

468'Transact. Ent. Soc.' 3rd series, vol. ii. ('Journal of Proceedings, p. 117.)

469Westwood, 'Modern Class. of Insects,' vol. i. p. 427; for crickets, p. 445.

470Mr. Ch. Horne, in 'Proc. Ent. Soc.' May 3, 1869, p. xii.

471The Oecanthus nivalis, Harris, 'Insects of New England,' 1842, p. 124.

472Platyblemnus: Westwood, 'Modern. Class.' vol. i. p. 447.

473B. D. Walsh, the Pseudo-neuroptera of Illinois, in 'Proc. Ent. Soc. of Philadelphia,' 1862, p. 361.

474'Modern Class.' vol. ii. p. 37.

475Walsh, ibid. p. 381. I am indebted to this naturalist for the following facts on Hetærina, Anax, and Gomphus.

476'Transact. Ent. Soc' vol. i. 1836, p. lxxxi.

477See abstract in the 'Zoological Record' for 1867, p. 450.

478Kirby and Spence, 'Introduct. to Entomology,' vol. ii. 1818, p. 35.

479See an interesting article, "The Writings of Fabre," in 'Nat. Hist. Review,' April, 1862, p. 122.

480'Journal of Proc. of Entomolog. Soc.' Sept. 7th, 1863, p. 169.

481P. Huber, 'Recherches sur les Mœurs des Fourmis,' 1810, p. 150, 165.

482'Proc. Entomolog. Soc. of Philadelphia,' 1866, p. 238-239.

483Quoted by Westwood, 'Modern Class. of Insects,' vol. ii. p. 214.

484Pyrodes pulcherrimus, in which the sexes differ conspicuously, has been described by Mr. Bates in 'Transact. Ent. Soc.' 1869, p. 50. I will specify the few other cases in which I have heard of a difference in colour between the sexes of beetles. Kirby and Spence ('Introduct. to Entomology,' vol. iii. p. 301) mention a Cantharis, Meloe, Rhagium, and the Leptura testacea; the male of the latter being testaceous, with a black thorax, and the female of a dull red all over. These two latter beetles belong to the Order of Longicorns. Messrs. R. Trimen and Waterhouse, junr., inform me of two Lamellicorns, viz., a Peritrichia and Trichius, the male of the latter being more obscurely coloured than the female. In Tillus elongatus the male is black, and the female always, as it is believed, of a dark blue colour with a red thorax. The male, also, of Orsodacna atra, as I hear from Mr. Walsh, is black, the female (the so-called O. ruficollis) having a rufous thorax.

485'Proc. Entomolog. Soc. of Philadelphia,' 1864, p. 228.

486Kirby and Spence, 'Introduct. Entomolog.' vol. iii. p. 300.

487Kirby and Spence, ibid. vol. iii. p. 329.

488'Modern Classification of Insects,' vol. i. p. 172. On the same page there is an account of Siagonium. In the British Museum I noticed one male specimen of Siagonium in an intermediate condition, so that the dimorphism is not strict.

489'The Malay Archipelago,' vol. ii. 1869, p. 276.

490'Entomological Magazine,' vol. i. 1833, p. 82. See also on the conflicts of this species, Kirby and Spence, ibid. vol. iii. p. 314; and Westwood, ibid. vol. i. p. 187.

491Quoted from Fischer, in 'Dict. Class. d'Hist. Nat.' tom. x. p. 324.

492'Ann. Soc. Entomolog. France,' 1866, as quoted in 'Journal of Travel,' by A. Murray, 1868, p. 135.

493Westwood, 'Modern Class.' vol. i. p. 184.

494Wollaston, On certain musical Curculionidæ, 'Annals and Mag. of Nat. Hist.' vol. vi. 1860, p. 14.

495'Zeitschrift für wiss. Zoolog.' B. xvii. 1867, s. 127.

496I am greatly indebted to Mr. G. R. Crotch for having sent me numerous prepared specimens of various beetles belonging to these three families and others, as well as for valuable information of all kinds. He believes that the power of stridulation in the Clythra has not been previously observed. I am also much indebted to Mr. E. W. Janson, for information and specimens. I may add that my son, Mr. F. Darwin, finds that Dermestes murinusstridulates, but he searched in vain for the apparatus. Scolytus has lately been described by Mr. Algen as a stridulator, in the 'Edinburgh Monthly Magazine,' 1869, Nov., p. 130.

497Schiödte, translated in 'Annals and Mag. of Nat. Hist.' vol. xx. 1867, p. 37.

498Westring has described (Kroyer, 'Naturhist. Tidskrift,' B. ii. 1848-49, p. 334) the stridulating organs in these two, as well as in other families. In the Carabidæ I have examined Elaphrus uliginosus and Blethisa multipunctata, sent to me by Mr. Crotch. In Blethisa the transverse ridges on the furrowed border of the abdominal segment do not come into play, as far as I could judge, in scraping the rasps on the elytra.

499I am indebted to Mr. Walsh, of Illinois, for having sent me extracts from Leconte's 'Introduction to Entomology,' p. 101, 143.

500M. P. de la Brulerie, as quoted in 'Journal of Travel,' A. Murray, vol. i. 1868, p. 135.

501Mr. Doubleday informs me that "the noise is produced by the insect raising itself on its legs as high as it can, and then sinking its thorax five or six times, in rapid succession, against the substance upon which it is sitting." For references on this subject see Landois, 'Zeitschrift für wissen. Zoolog.' B. xvii. s. 131. Olivier says (as quoted by Kirby and Spence, 'Introduct.' vol. ii. p. 395) that the female of Pimelia striata produces a rather loud sound by striking her abdomen against any hard substance, "and that the male, obedient to this call, soon attends her and they pair."

502Apatura Iris: 'The Entomologist's Weekly Intelligencer,' 1859, p. 139. For the Bornean Butterflies see C. Collingwood, 'Rambles of a Naturalist,' 1868, p. 183.

503See my 'Journal of Researches,' 1845, p. 33. Mr. Doubleday has detected ('Proc. Ent. Soc.' March 3rd, 1845, p. 123) a peculiar membranous sac at the base of the front wings, which is probably connected with the production of the sound.

504See also Mr. Bates' paper in 'Proc. Ent. Soc. of Philadelphia,' 1865, p. 206. Also Mr. Wallace on the same subject, in regard to Diadema, in 'Transact. Entomolog. Soc. of London,' 1869, p. 278.

505'The Naturalist on the Amazons,' vol. i. 1863, p. 19.

506See the interesting article in the 'Westminster Review,' July, 1867, p. 10. A woodcut of the Kallima is given by Mr. Wallace in Hardwicke's 'Science Gossip,' Sept. 1867, p. 196.

507See the interesting observations by Mr. T. W. Wood, 'The Student,' Sept. 1868, p. 81.

508Mr. Wallace in 'Hardwicke's Science Gossip,' Sept. 1867, p. 193.

509See also, on this subject, Mr. Weir's paper in 'Transact. Ent. Soc.' 1869, p. 23.

510'Westminster Review,' July, 1867, p. 16.

511For instance, Lithosia; but Prof. Westwood ('Modern Class. of Insects,' vol. ii. p. 390) seems surprised at this case. On the relative colours of diurnal and nocturnal Lepidoptera, see ibid. p. 333 and 392; also Harris, 'Treatise on the Insects of New England,' 1842, p. 315.

512Such differences between the upper and lower surfaces of the wings of several species of Papilio, may be seen in the beautiful plates to Mr. Wallace's Memoir on the Papilionidæ of the Malayan Region, in 'Transact. Linn. Soc.' vol. xxv. part i. 1865.

513'Proc. Ent. Soc.' March 2nd, 1868.

514See also an account of the S. American genus Erateina (one of the Geometræ) in 'Transact. Ent. Soc.' new series, vol. v. pl. xv. and xvi.

515'Proc. Ent. Soc. of London,' July 6, 1868, p. xxvii.

516Harris, 'Treatise,' &c., edited by Flint, 1862, p. 395.

517For instance, I observe in my son's cabinet that the males are darker than the females in the Lasiocampa quercus, Odonestis potatoria, Hypogymna dispar, Dasychira pudibunda, and Cycnia mendica. In this latter species the difference in colour between the two sexes is strongly marked; and Mr. Wallace informs me that we here have, as he believes, an instance of protective mimickry confined to one sex, as will hereafter be more fully explained. The white female of the Cycnia resembles the very common Spilosoma menthrasti, both sexes of which are white; and Mr. Stainton observed that this latter moth was rejected with utter disgust by a whole brood of young turkeys, which were fond of eating other moths; so that if the Cycnia was commonly mistaken by British birds for the Spilosoma, it would escape being devoured, and its white deceptive colour would thus be highly beneficial.

518'Rambles of a Naturalist in the Chinese Seas,' 1868, p. 182.

519Wallace on the Papilionidæ of the Malayan Region, in 'Transact. Linn. Soc. vol. xxv. 1865, p. 8, 36. A striking case of a rare variety, strictly intermediate between two other well-marked female varieties, is given by Mr. Wallace. See also Mr. Bates, in 'Proc. Entomolog. Soc.' Nov. 19th, 1866, p. xl.

520Mr. R. MacLachlan, 'Transact. Ent. Soc.' vol. ii. part 6th, 3rd series, 1866, p. 459.

521H. W. Bates, 'The Naturalist on the Amazons,' vol. ii. 1863, p. 228. A. R. Wallace, in 'Transact. Linn. Soc.' vol. xxv. 1865, p. 10.

522On this whole subject see 'The Variation of Animals and Plants under Domestication,' vol. ii. 1868, chap. xxiii.

523A. R. Wallace, in 'The Journal of Travel,' vol. i. 1868, p. 88. 'Westminster Review,' July, 1857, p. 37. See also Messrs. Wallace and Bates in 'Proc. Ent. Soc.' Nov. 19th, 1866, p. xxxix.

524'The Variation of Animals and Plants under Domestication,' vol. ii. chap. xii. p. 17.

525'Transact. Linn. Soc.' vol. xxiii. 1862, p. 495.

526'Proc. Ent. Soc.' Dec. 3rd, 1866, p. xlv.

527'Transact. Linn. Soc.' vol. xxv. 1865, p. 1; also 'Transact. Ent. Soc.' vol. iv. (3rd series), 1867, p. 301.

528See an ingenious article entitled, "Difficulties of the Theory of Natural Selection," in the 'Month,' 1869. The writer strangely supposes that I attribute the variations in colour of the Lepidoptera, by which certain species belonging to distinct families have come to resemble others, to reversion to a common progenitor; but there is no more reason to attribute these variations to reversion than in the case of any ordinary variation.

529Wallace, "Notes on Eastern Butterflies," 'Transact. Ent. Soc.' 1869, p. 287.

530Wallace, in 'Westminster Review,' July, 1867, p. 37; and in 'Journal of Travel and Nat. Hist.' vol. i. 1868, p. 88.

531See remarks by Messrs. Bates and Wallace, in 'Proc. Ent. Soc.' Nov. 19, 1866, p. xxxix.

532See Mr. Wallace in 'Westminster Review,' July, 1867, p. 11 and 37. The male of no butterfly, as Mr. Wallace informs me, is known to differ in colour, as a protection, from the female; and he asks me how I can explain this fact on the principle that one sex alone has varied and has transmitted its variations exclusively to the same sex, without the aid of selection to check the variations being inherited by the other sex. No doubt if it could be shewn that the females of very many species had been rendered beautiful through protective mimickry, but that this has never occurred with the males, it would be a serious difficulty. But the number of cases as yet known hardly suffices for a fair judgment. We can see that the males, from having the power of flying more swiftly, and thus escaping danger, would not be so likely as the females to have had their colours modified for the sake of protection; but this would not in the least have interfered with their receiving protective colours through inheritance from the females. In the second place, it is probable that sexual selection would actually tend to prevent a beautiful male from becoming obscure, for the less brilliant individuals would be less attractive to the females. Supposing that the beauty of the male of any species had been mainly acquired through sexual selection, yet if this beauty likewise served as a protection, the acquisition would have been aided by natural selection. But it would be quite beyond our power to distinguish between the two processes of sexual and ordinary selection. Hence it is not likely that we should be able to adduce cases of the males having been rendered brilliant exclusively through protective mimickry, though this is comparatively easy with the females, which have rarely or never been rendered beautiful, as far as we can judge, for the sake of sexual attraction, although they have often received beauty through inheritance from their male parents.

533'Proc. Entomolog. Soc.' Dec. 3rd, 1866, p. xlv., and March 4th, 1867, p. lxxx.

534See Mr. J. Jenner Weir's paper on insects and insectivorous birds, in 'Transact. Ent. Soc.' 1869, p. 21; also Mr. Butler's paper, ibid. p. 27.

22322477R00112

Made in the USA
San Bernardino, CA
30 June 2015